Philip Henry Zwicker

Zwicker's Revised

Practical instructor in questions and answers for machinists, firemen,

electricians, and steam engineers

Philip Henry Zwicker

Zwicker's Revised
Practical instructor in questions and answers for machinists, firemen, electricians, and steam engineers

ISBN/EAN: 9783337254490

Printed in Europe, USA, Canada, Australia, Japan

Cover: Foto ©berggeist007 / pixelio.de

More available books at **www.hansebooks.com**

ZWICKER'S

REVISED.

PRACTICAL

INSTRUCTOR

IN

QUESTIONS AND ANSWERS

FOR

Machinists, Firemen, Electricians

AND

STEAM ENGINEERS.

BY

PHILIP HENRY ZWICKER,

Practical Engineer and Machinist.

1607 Wash Street.

ST. LOUIS, MO.
1890.

INTRODUCTION.

This book is written for the special information of Engineers, Machinists, Firemen and Electricians who have to sooner or later procure an engineers license by going before a board of practical engineers and answering questions relating to the care of boilers, pumps, injectors, engines, indicator safety-valve and electric light dynamo. Before they can collect their salary as an engineer. It is hoped its practical suggestions will enable those who follow them to gain a better insight of the work they have to perform. The real object in writing this book is to help my fellow man and not keep him in the dark. All questions are answered in plain and simple language, so any man of limited education can thoroughly understand.

Very truly yours,

P. H. ZWICKER, Author.

ZWICKER'S

REVISED

PRACTICAL INSTRUCTIONS

—IN—

QUESTIONS AND ANSWERS

—FOR—

Machinists, Firemen, Electricians and Steam Engineers

Q. What are the duties of an Engineer?

A. His duties are to take full charge of the boilers and engines where ever he may be employed, and see that the steam machinery under his charge are kept in No. 1 order with little expense to his employer.

Q. What is required of a man to become a first-class Engineer?

A. He is obliged to obtain an Engineers license touching his qualifications as an engineer of steam engines, by which will be shown that he is a suitable and safe person to be intrusted with tle powers and duties of an engineer.

Q. What experience must a man have in order to get his application before a board of Engineers?

A. His experience must be generally two years, at mechanical or steam engineering, which must be sworn to by two citizens, one being a licensed engineer and the other a good reliable citizen, both living in the city, where applicant has worked.

QUESTIONS AND ANSWERS.

Q. What business do you follow?

A. ....

Q. What is a steam boiler, how is it, and of what is it made?

A. A steam boiler is a closed vessel made of steel, iron or copper plates, the most in use is $\frac{3}{8}$ $\frac{1}{2}$ and $\frac{5}{16}$ inch thick, and ranging from 45,000 to 85,000 lbs. tensile strength; these plates are run through a rolling machine and rolled in a circle, then riveted together, generally with two rows of rivets, because the strain is greater side-wise than endwise, the seams around the boiler are single riveted because the strain is not so great; the boiler is braced by different kinds of braces, such as a crow foot, Longitudinal dome, side braces, etc. The eye is riveted to the head of the boiler, which head is generally made of $\frac{5}{8}$

inch plate, the other eye is riveted to the side top or dome of boiler, and the brace and eye are put together by bolts with a split key to keep the belt in its place.

Q. Which are the chief points in the construction of a successful and economical boiler?

A. Proper circulation facilities constitute one of the chief points in the construction of a successful and economical boiler. In tubular boilers, the best practise is to place the tubes in vertical rows, (plumb) leaving out what would be the center row. The circulation is up the sides of the boiler and down the center. Tubes set zigzag or to break spaces impede the circulation and will not practically give the best results.

Q. How should a brace fit?

A. It should fit tight, for if it were loose it would be of no account.

Q. If you found a brace loose, what would you do, and how would you tighten it

A. By taking the brace out, heat it in the center, then upset it by jumping it endwise on a block of wood until it is the proper length.

Q. Why is a boiler braced?

A. For strength.

Q. What is a stay bolt?

A. A stay bolt is a screw bolt, put through an outside and into an inside sheet, so as to hold them that they may not spread or collapse, such as a fire-box sheet and an outside shell, they are put together with stay bolts so as to allow a water space between the two sheets.

Q. How are stay bolts made and put in?

A. They are made with one continuous thread, and screwed through the outside, then through the space between, then through the fire-box sheet and allowed to stick through $\frac{3}{16}$ of an inch, so they can be riveted over each end to act as a brace, the space between the two sheets is called a water space.

Q. What is meant by corrosion?

A. It means wasting away of the iron of boilers plates by pitting, grooving, etc. There is internal and external corrosion; the acids in and minerals in the water liberated by the heat, attack the boiler internally, and the sulphur which comes out of the coal has a strong attachment for iron, and that attacks the outside.

Q. How would you find the water level when your boiler is foaming?

A. The proper way would be to shut down the engine and all valves connected with the boiler, cover fire with ashes, close the damper, then the water will quiet down, and the level of the water easily found. An engineer should know when lighting a fresh fire, never to force it, but let it heat gradually, so that all parts expand as near equal as possible; good judgment is needed. Boilers and steam guages should be tested at least once a year.

Q. Where would you put a steam gauge?

A. Sometimes on top of the boiler, and in some cases on the steam drum. It must always be tapped into the steam part of the boiler, the shorter the pipe the better. The steam gauge and safety valve should correspond under all circumstances.

Q. Why is a pet cock put under the steam guage?

A. To drain the pipe in cold weather.

Q. What kind of a steam gauge have you got? A. A spring gauge.

Q. What is a steam gauge for?

A. To indicate the pressure in pounds per square inch in the boiler.

Q. Does the steam gauge get out of order?

A. Yes, sometimes.

Q. If the steam gauge was out of order what would you be governed by?

A. By the safety valve.

Q. How would you know that it was in working order?

A. By raising the lever two or three times to see that the valve is not stuck.

Q. What is a safety valve for?

A. It is intended to release the boiler and prevent explosions from over-pressure.

Q. How large should the safety valve be in proportion to the boiler, and grate surface?

A. The safety valve should be $\frac{1}{2}$ square inch to each square foot of grate surface, which will make it large enough to relieve the boiler of all steam generated over which the safety valve is set.

Q. Which are the better, gauge cocks or glass gauges, and which would you be governed by?

A. Gauge cocks, because glass gauges are liable to get stopped with mud, and not give a true level of the water, but they are a very handy thing; they should be blown out four or five times a day, so as to keep them from clogging up.

Q. What would you do in case a glass should happen to break?

A. First close the water valve to prevent the escape of water, close the steam valve, insert a new glass, then turn on the steam valve first, the water valve next, then close the pet cock at the bottom and everything will be all right.

Q. What is the best way to clean a glass gauge inside?

A. The best way, is to take a small piece of waste and tie it to a strong thin stick, saturate the waste with soap or acetic acid, pass down inside of the glass, then blow through with steam and the glass will be clean as new. Never touch the inside of a glass water gauge with wire, if you do, it will crack. The best glasses are the Scotch brand, called Eureka.

Q. If your gauge cock, or a small pipe in

the large steam pipe, should happen to get broken off, what would you do?

A. Make a hard wood plug and drive it in with a heavy hammer, then leave it so until it could be repaired, by cutting out the old piece, retapping and putting in another pipe or gauge cock, whichever the case may be.

Q. `What clearance should a boiler have?

A. It should have from 3 to 4 inches at the fire-line, and from 5 to 7 inches between the shell and bridge wall; a boiler should have from 2 to 3 bridge walls so the fire will hug the boiler; it also makes the coal burn cleaner and steams easier. The first bridge wall should be on the back end of the grate bars, and others about 3 to 5 feet apart, according to the length of the boiler. Where the smoke returns through the flues, it should be about $\frac{1}{5}$ larger than the area of flues or tubes combined, bridge walls should lean toward the back.

Q. How should a boiler rest and what on?

A. The front end of the boiler should rest on the fire front, and the back end generally rests on a cast iron leg or two or three rollers, to allow the boiler to expand equally. The mud drum should always hang free under all circumstances.

Q. In what should engineers be careful and exercise good judgment?

A. Engineers should be careful in starting or stopping an engine with a high pressure of steam.

Q. Why should engineers be careful in starting or stopping an engine?

A. Because the rent in giving the steam in starting, and the sudden check in stoping, may cause such a pressure as to rupture the boiler.

Q. What else should engineers look after?

A. Engineers should see that their draft is not choked by ashes under the boiler, and that the outside of the boiler and inside of flues are kept clean, then they will have no trouble in keeping up steam.

Q. How would you clean the flues or tubes of a steam boiler?

A. By either blowing steam through them or using a flue cleaning brush.

Q. How are flues or tubes cleaned by steam?

A. Some boilers have a 1½ inch pipe with a valve attached, also branch pipes of smaller dimensions, leading from the 1½ inch pipe into the back end and into the flues; others have a hose attached to the front end leading from the steam drum, so the flues or tubes can be blown out from the front end. (Cleaning by the brush is the better and more popular way.)

Q. How often would you clean out the flues and when?

A. Once a day, in the afternoon, sometimes in the morning after raising steam.

Q. What different strains has a boiler?

A. To the flues or tubes it has a crushing strain, to the shell a tearing strain.

Q. What causes boiler explosions?

A. There are various causes, such as low water, over-pressure of steam, bad safety valve, foaming boilers and burnt sheets.

Q. Why would a foaming boiler cause an explosion?

A. It generally raises the water from the heated sheets. They become hot; the water falling back on them they crack, and some–

times cause an explosion. A blistered sheet or a scaly boiler will also cause an explosion, by allowing the sheets to become burnt and weakened; also an untrue steam gauge is very bad.

Q. What are the worst explosions?

A. The worst explosions are caused by high pressure and plenty of water; low water allows the iron to burn and crack, which weakens it, and when the cold water touches it, it does not take so much to burst.

Q. How would you know if your boiler had blistered sheets or was rotten?

A. By the hammer test; by taking a small hammer and going inside and outside of the boiler and seeing if it is all right by sound.

Q. How would you know by sound?

A. By the different sounds it has; if it rings and sounds solid it is all right; but if it sounds dead, hollow or blunt, there is something wrong.

Q. Would you strike the iron hard?

A. Yes, pretty hard.

Do not hesitate to have a boiler insured, as insurance is generally accompanied by hammer test and intelligent inspection, which guarantees safety to the engineer or owner.

Do not reject the advice or suggestions of intelligent boiler inspectors, as their experience enables them to discriminate in cases which never come under the observation of men who do not follow inspection as a business.

Q. If you wished to put a patch on a boiler what kind would you put on?

A. A hard patch; it is reliable and safe.

Q. Why not put on a soft patch?

A. Because they are not reliable and are dangerous.

Q. What is the difference between a hard and soft patch ?

A. A hard patch is a patch where the piece is cut out of the boiler and rivet holes are drilled or punched through, then the patch is riveted on, chipped, caulked and made water and steam tight.

Q. What is a soft patch ?

A. A soft patch is put over the plate that needs patching, and put on with $\frac{5}{8}$ or $\frac{3}{4}$ inch countersunk screw bolts, and a mixture of red lead and iron borings to put between the patch and boiler; the piece of sheet in the boiler is not cut out for a soft patch as in a hard patch, consequently the patch is burnt, as the water in the boiler can not come in contact with the patch.

Q. Which are the better, drilled or punched holes? A. Drilled holes.

Q. Why?

A. Because the fiber of the iron is not disturbed as in punching; in drilling, the iron is cut out regular; in punching, it is forced out at once.

Q. What should be the proper rivets for certain sized sheets, and how far apart?

A. The rivets should be $\frac{5}{8}$ and $\frac{3}{4}$ inch diameter, and $1\frac{3}{4}$ to 2 inches apart.

Q. Before shutting down at night, what should be done?

A. Pull out the fire, pump up to the third gauge and close the glass gauge cocks, so that in case the glass should happen to get broken during the night, the water could not escape.

Q. What would you do the first thing

in the morning on entering the boiler-room.

A. See how much water was in the boiler by trying the gauge cocks, then open the glass gauge valves, and start the fire to raise steam.

Q. Why do you try the gauge cocks, and not trust to the glass gauge?

A. Because the water pipe connecting the glass gauge with the boiler is liable to become stopped up with mud, consequently the glass would not show a true level of water. The glass gauge should be blown out five or six times a day, to insure safety, but never depend on the glass gauge alone.

Q. If you found too much water in the boiler during the day, what would you do?

A. Open the blow-off valve and let out water to the second gauge. An engineer should be very careful when blowing out water when he has a hot fire in the boiler furnace, as the water leaves very fast, and may blow out too much; good judgment should be used.

Q. How would you clean a boiler?

A. First see that there is no fire under the boiler, then let out all the water through the blow-off valve, take out the man, hand, and mud-drum plates; then take a short-handle broom, a candle or torch, a small hand-pick, a scraper made out of an old file flattened on the end and bent to suit, also a half-inch square iron twisted link chain, about 3 feet long, with a ring at each end to answer for a handle; place the chain around the flue and work the chain to get the scale off the bottom of the flues; use the pick and scraper to pick and scrape off all that can be seen on top of the flues and the bottom and sides of the shell; then wash out into the mud-drum; clean out and put in the mud-drum and hand-hole plates; fill up to top of flues; then put in the man-hole plate, and fill up to the second gauge ready for raising steam.

Q. Could a boiler not be blown out?

A. Yes, but not practically.

Q. How much pressure would you allow?

A. About 10 or 20 pounds.

Q. Why not more pressure?

A. Because the heat would be so great that the expansion and contraction would not be equal; consequently, the boiler seams would probably leak and the boiler be injured. The better way is no steam pressure.

Q. What benefit is gained by letting the water stay in the boiler until you are ready to clean it out?

A. The mud is kept soft and the scale is not caked to the shell or tubes; also, the seams and the boiler are not injured by unequal expansion and contraction.

Q. How should man and hand-hold plates be taken out and put in?

A. They should be marked with a chisel at the top, also the boiler at man-hole and hand-hole, whichever it might be, and they should be put in the same way they came out.

Q. How would you gasket the man-hole or hand-hole plates of a boiler?

A. With pure lead rings; some use sheet rubber, etc.

Q. Why are man-hole and hand-hole plates made oblong instead of round?

A. Because if they were round they could not be taken out or put in, and a man could not easily enter the boiler.

Q. When filling a boiler with cold water, and raising steam, what should be done?

A. A valve should be left open.

Q. Why do you leave a valve open?

A. Because a boiler fills easier and quicker, and in raising steam the cold air is let out, which allows equal expansion, as cold air prevents equal expansion.

Q. How would you set a boiler?

A. By using a spirit level across and along the flues, allowing the end furthest from the gauge cocks $\frac{1}{4}$ inch lower for every 10 feet in length. Q. Why?

A. Because when there is water in the gauge cocks, there will surely be water in the other end.

Q. How many gauge cocks has a boiler?

A. Generally three.

Q. Where is the first?

A. Two inches above the flues, and the rest two inches apart.

Q. Where is the water line? A. First gauge.

Q. Where would you carry water when running? A. Second gauge.

Q. Where would you carry water when shutting down at-night? A. Third gauge.

Q. Why?

A. To allow for evaporation and leakage.

Q. Where is the fire line of a boiler?

A. In line or little below first gauge.

Q. When you open a boiler and look in, where do the scales form and lay thickest?

A. Over the fire-plates and around the mud-drum leg or blow-off pipe.

Q. Why?

A. Because the circulation and heat are greatest there.

Q. What is a steam drum for?

A. To have more steam in volume.

Q. Which is the hotter, coming out of the same boiler, steam or water?

A. They are the same, only water will retain the heat longer, as water is a fluid and steam a vapor.

Q. How should the circulation and feed be?

A. The circulation and feed should be continual. Q. Why?

A. Because boilers have exploded just as

the steam valve was opened to start the engine, after having stood still for some time. This is generally caused by the plates that are in con-tact with the fire becoming overheated, as the circulation being stopped after the steam is shut off. And just as soon as the valve is opened the pressure becomes lessened, and the water on the overheated sheets flashes into steam of GREAT ELASTIC FORCE, and if the boiler is not strong enough, a terrific explosion is the result.

Q. If you tried the gauge cocks and found no water in sight, what would you do?

A. Simply shovel wet ashes over the fire pull it out, raise the flue caps and let the boiler cool down.

Q. Why do you throw wet ashes over the fire before pulling it out?

A. If the fire was stirred up it would create more heat and be liable to burn the plates.

The braces in the boiler should be examined to see if they are loose, also the sheets, flues, heads and seams, to see if they are cracked or leaking ; if they are not attended to, they may cause trouble and loss of life and limb. Engineers should not allow anything about the engine or boiler room to become greasy or dirty, for it shows poor management, and a careless, worthless engineer. If valves or cocks leak, they should be ground in with emery and oil until a seat or true bearing is found. Ground glass is good for grinding brass valves.

Q. When should the boiler seams be caulked ?

A. When the boiler is empty and cold, for when the boiler is hot and filled with water, the jarring while caulking would have a tendency to spring a leak somewhere else.

Q. Would you call pressure and weight the same ? A. No. Q. Why?

A. Because pressure forces in every direction, while weight presses down.

Q. Which is best, the riveted or the lap-welded flues ?

A. The lap-welded flues, as they are a true circle and not so easily collapsed as the riveted flues. Q. Why ?

A. Because the riveted flues are not a true circle.

Q. What is foaming?

A. Foaming is the water and steam mixed together.

Q, What causes foaming?

A. Dirty, greasy, oily and soapy water; salt water forced into fresh water, also too much water and not enough steam room, will cause foaming.

Q. What is priming?

A. Priming is the lifting of water with steam, such as opening a valve suddenly, and drawing the water from the boiler to the cylinder of the engine.

Q. What would you do in that case?

A. Close the throttle valve and leave it closed for a few minutes, then open the valve slowly; that will remedy it. Sometimes priming is caused by too much water and not enough steam room; in that case carry less water.

Q. Are boilers sometimes injured by the hydraulic test?

A. Yes, if tested by an inexperienced person. The hydraulic test is the safest, because if the boiler is bursted no one is likely to get hurt. Never use steam pressure under any circumstances for testing.

Q. If you had a high pressure of steam, and water was out of sight, would you raise the safety valve to let off the pressure?

A. No, under no circumstances.

Q. Why not?

A. Because it would cause the water to rise, and when the valve closed the water would drop on the heated parts and be liable to cause an explosion.

Q. If your boiler was too small to keep up the amount of steam required, would you weight down the safety valve to carry a higher pressure? A. No.

Q. Why not?

A. Because it would show carelessness and a violation of the laws. There is no mystery about boiler explosions. They are simply caused by carelessness, and no man has the right to endanger the lives and property of others when he knows that he is incompetent to perform the duty required of him as an engineer, whether licensed or otherwise.

Q. How much space should there be between the tubes of a steam boiler?

A. The space should be one-half the diameter of the tube itself.

Q. Can you name the principal valve on a steam boiler?

A. Yes. The safety valve, by all means.

Q. Where should the lower gauge cock be placed in an upright boiler, any size boiler?

A. One-third the distance from the top, between the two flue sheets.

Q. How long a time would you consider it safe to leave the engine room alone without attention?

A. Under no circumstances should the engine or boiler room be left alone.

Q. Why not, when everything is in working order? A. Because no man can tell at what moment an accident might occur, which if neglected might cause a serious loss of life and property.

Q. What is the boiling point of water?

A. It is 212 degrees of heat.

Q. At what point does water turn into steam? A. It evaporates at 213 degrees.

PUMPS.

Q. What kinds of pumps are there?

A. There are many kinds, but we consider only single action and double action for feeding boilers and general use.

Q. How many valves has a single action plunger pump?

A. Two valves, a receiving valve and a discharge valve.

Q. How many valves has a double action?

A. Four, two receiving and two discharging. The double action receives and discharges both strokes. This kind of pump has a steam cylinder on one end. Large pumps have eight, sixteen and thirty-two small valves on water cylinder, according to the size of the pump.

Q. Why do large pumps have many small water valves and not a few larger ones in proportion?

A. The reason the pumps have small valves is that the valves do not have to open as much as larger ones, consequently the pump does not loose the quantity of water each stroke as it would with larger valves.

Q. How are pumps set up and leveled?

A. Set the pump so the receiving is from the boiler and the discharge toward the boiler, put in the same size receiving and discharge pipe as tapped in the pump, so the pump can have a good supply and discharge. The pump is leveled with a spirit-level or a square and plumb line. To level a double-action pump, some level across the frame and along the piston; the other way is to take the valve chamber cap off the water cylinder and level the valve seats, so the valves will rise and drop plumb. To level a single action pump, take off the valve chamber caps and level both ways.

Q. How are the steam valves of duplex pumps set and adjusted?

A. Take off the valve chest cover, shove the piston against one of the cylinder heads and mark the piston rod with a pencil at the packing-box gland, then shove the piston against the other cylinder-head and make a another mark, find the center between the two marks and move the piston until the center mark reaches the packing-box gland where the first mark was made. Or in other words plumb the lever that connects the valve rocker shaft and the piston. After this is done, see how the steam valve is for lead; if equal at both ends the valve is set, if not, adjust by uncoupling the valve stem at the coupling outside of the packing box, and turn to suit the adjustment in equalizing the "lead."

Q. How is the water piston packed and with what in the water cylinder?

A. It is generally packed with square canvas and rubber mixed packing; it generally takes two pieces; one piece is jointed on top, and the other at the bottom, to make what engineers call a broken joint. The packing runs from $\frac{1}{4}$ to $\frac{3}{8}$ inch square. These are the general sizes used for common sized pumps.

Q. What other valve has a pump near the boiler?

A. A check valve.

Q. What is a check valve for?

A. To check the water in the boiler from

coming back, in case there is any work to be done on the pump.

Q. Could you pump water into the boiler if you had four or five check valves on the discharge pipe? A. Yes, I could force through all, but it would be more labor on the pump, because the plunger would have to force harder to raise the number of check valves.

Q. Where is a pet cock put on the pump barrel for cold water, and why?

A. It is put at the side and near the bottom of the pump barrel, and is there to show how the pump is working, and to drain the pump in winter to prevent freezing.

Q. How do you know when your pump is in good working order?

A. By opening the pet cock and seeing the stream that comes out.

Q. How does it show when the pump is in good working order ?

A. Nothing in the suction stroke and full force in the discharge stroke.

Q. Where would you locate the trouble if it came full force both strokes?

A. I would locate it at check and discharge valves, both being caught up.

Q. Where would you locate the trouble if it came full force both strokes, moderate, tank or hydrant pressure?

A. At the receiving valve.

Q. Can you run a pump without a check valve?

A. If the discharge valve is in good order, it can; but if there is neither check nor discharge, it can not.

Q. Can you feed a boiler without a pump?

A. If the pressure of the boiler is below the pressure of the feed water or city pressure, it can, by simply opening a water valve and letting in the amount of water required.

Q. What other way is a boiler fed?

A. By an injector or an inspirator.

Q. What is an injector or an inspirator?

A. They are devices to answer for a pump in feeding a boiler; they draw force and heat the water at the same time. See Page 46.

Q. Must a pump have a valve?

A. Yes, if a pump had no valve it would not

do any work. A pump is not a pump unless it has a valve. There are common well hand pumps with one valve, called a receiving or suction valve, but a force pump has two valves,. a receiving and discharge ; the discharge is to retain the water after it is delivered, so the plunger can get a fresh supply. After the plunger has ascended and begins to descend, the water sets on top of the receiving and under the discharge; consequently, when the plunger descends it forces the receiving shut and the discharge open.

Q. Should there not be another valve near the boiler ?

A. Yes, a globe valve between the check valve and the boiler.

Q. What is that for ?

A· To close and keep the pressure in the boiler in case the check valve is caught up and needs repairing.

Q. Can you raise, lift or suck hot water with a pump ? A. Not very well.

Q. Why ?

A. Because the pump would get steam

bound. Hot water should be level o. higher than the pump in order to work well.

Q. Where should a pet cock be put on the pump barrel for hot water ?

A. At the top of barrel, immediately under the packing ring.

Q. Why is it put there ?

A. To let out steam when steam bound, and air when air bound. There should be a pet cock tapped in the cap of the valve chamber to let off the steam or air when steam or air bound.

Q. If you had no pet cock on the valve chamber cap, what would you do ?

A. Simply take a wrench and loosen one of the nuts a little until the air or steam was out, then tighten it again.

Q. Why is an air chamber put on a double action pump, and what is it?

A. It is simply a copper vessel air tight. When the pump is working, the water is forced up into the chamber, compresses the air, and the air acts as a cushion on the valves and piston head in the water cylinder.

Q. What is a cushion ?

A. A cushion is anything that is compressed, and by its compression is formed into a higher and stronger pressure, consequently acting as a spring, deadening any knock that might have occurred otherwise, as water will cause a knock, it being nearly as solid as iron, so if a double action pump had no air chamber, there would be a continual thumping noise. Q. What is a vacuum ?

A. A vacuum is a space void of matter.

Q. Can a perfect vacuum be formed ?

A. No, about 9 to 11 per cent. of the atmosphere, which is 14.7 pounds per square inch.

Q. What will a vacuum do ?

A. It will lift water 33 feet, providing all pipes and connections are air tight.

Q. How is a vacuum created or made ?

A. When the plunger of a pump is well packed and it lifts, it excludes the air out of the pump barrel and suction pipe, consequently the water, being at the other end of the pipe, it follows the plunger; or, in other words, the atmospheric pressure, being 14.7

pounds per square inch, forces the water up the pipe to fill the vacancy made by the plunger forming the vacuum. Q. What should be placed at the bottom of the suction pipe ?

A. A strainer made out of gauze wire, a foot valve and a pet cock to drain it.

Q. If your pump should not be working, your water running low, and you were asked to run a little while longer, would you run and let your water become dangerously low ?

A. No, take no chances whatever, but shut down and go about repairing the trouble.

Q. Where would you look for the trouble ?

A. Open the pet cock of the pump, and that will very nearly tell where to look for it; if no water comes out, the water is shut off, or there is none, etc.

Q. What generally prevents a pump from working ?

A. Not enough water, too small a suction pipe and obstruction of the valves to seat, by straws, sticks or anything that may be drawn through the suction pipe, or the pump valves becoming hot and sticking.

Q. If an accident happened, such as a broken pipe connected with the boiler and pump, or you could not get sufficient water to supply the boiler, what would you do?

A. Simply shut down the engine and all valves connected with the boiler, draw fire, raise flue caps, and close the damper, so as to keep what water there is in the boiler until the difficulty is repaired.

Q. If your suction pipe should spring a leak, what would you do?

A. Take a piece of sheet rubber, some copper wire, wrap around tight, and stop the leak temporarily. Q. If your hydrant, that supplies pump with water, should happen to get broken, what would you do?

A. First see how much water was in the boiler, by trying gauge-cocks, then shut off the water in the street, or wherever the lazy cock lay, and try to wrap it, if possible, or repair it. If an injector or inspirator was attached, and was supplied from a tank or well, use either.

Q. For instance, if you had neither of these, what would you do?

A. Shut down the engine, close the damper, raise the flue caps and draw the fire, whichever suited the circumstances.

Q. If your pump was turned around, could you feed the boiler? A. No.

Q. What would be the consequence?

A. If the packing in the pump held out, the plunger would exclude the air and collapse the discharge pipe.

Q. Would it not have a tendency to drain the water out of the boiler?

A. No, the check valve near the boiler would keep it back.

Q. If you had no check valve, what would it do?

A. The water would run out, that is, pro viding the pump was turned around.

Q. If the pump plunger is one-half the stroke of the engine, what should the diameter of the plunger be?

A. One-third the diameter of engine cylinder.

Q. How high should a valve lift to clear itself?

A. About one-fourth of its diametor or one-third of its area.

Q. What proportions should the valves be to any sized pump?

A. They should be one-fourth the area of the pump.

Q. Suppose in the evening when you shut down, that the pump was in good working order, and when you started up the next morning and opened the pump pet cock a strong stream of water came out both strokes: where would you locate the trouble?

A. The trouble would be at both the check and discharge valves being caught up.

Q. Suppose you started the pump and it was in good order, and no water came; where would you locate the trouble?

A. The suction pipe is leaking, or it is out of water, or there is no water.

Q. State the usual area proportion of the cylinders of a steam pump?

A. The steam cylinder averages four times the area of the water cylinder

THE ENGINE.

Q. What is a steam engine?

A. A steam engine is a machine by which power is obtained from steam.

Q. What is steam?

A. Steam is a gaseous vapor from water, generated by heat, composed of hydrogen and oxygen.

Q. How do you know water is composed of hydrogen and oxygen?

A. Science shows that 1 pound of hydrogen with 8 pounds of oxygen is equal to 9 pounds of water.

Q. What is an engine composed of?

A. A bed plate, cylinder, connecting rod, crank, crank-shaft, main pillow block, out pillow block, cross-head, wrist-pin in cross-head, crank-pin, two cylinder heads, piston-rod, piston-head, follower head, bull-ring, packing-rings, follower plate and bolts, connecting rod and brasses, pillow-block brasses, a valve, and guides where the cross-head slides in, so the piston is kept central with the cylinder. The main pillow-block brasses are generally made into four pieces, called top, bottom and two quarter brasses four sides of shaft; they are made into four parts, so as to take up lost motion.

Q. What keeps the rod from running off the crank pin?

A. The shoulder on the crank-pin.

Q. Why are the stub ends of straps made heavier where the gib and key pass through?

A. To make up for the amount of iron taken out for the gib and key-way.

Q. If water should accumulate in the cylinder, what would be the consequence?

A. It is liable to crack the cylinder and disable the engine.

Q. If you had charge of an engine in the country, and the cylinder head should happen to crack, how would you remedy it?

A. If not broken too bad, try to patch it with pieces of iron or boards, and brace it from the wall with a piece of heavy scantling, then try and run the engine until a new cylinder head could be made.

Q. What size should a steam pipe and an exhaust pipe be to any size cylinder ?

A. The steam pipe should be one-fourth and the exhaust pipe one-third the diameter of the engine cylinder itself.

Q. If your crank pin or other journals became hot, what would you do ?

A. Try, while running, to get water on them, then oil them; if that would not do, stop and slack up the key a little, then start up again. All engine cylinders should be well drained and heated before starting, then the engine should be started slowly, as the water that accumulates in the cylinder may injure the piston, cylinder, or cylinder heads. Always leave the cylinder cocks open when not running, and they should remain so until the cylinder is heated by the steam,—after the engine has been running at full speed two or three minutes at least.

Q. If the cylinder had shoulders inside, and was out of a true circle, what would you do to remedy it ?   A.   Bore it, or have it bored out.

Q.   In case, the throttle valve should become loose from the stem and prevent the steam from entering the valve chest, what would you do to repair it ?

A.   Close the valve next to the boiler, if there was one; if not, let the boiler cool down, then take the valve out and repair it.

Q.   If your side-valve was not steam-tight, what would you do ?

A.   Have the valve planed, then chip, file and scrape the seat to a full bearing.

Q.   If the crank and wrist-pins are worn out of true, what would you do ?

A.   Caliper and file them until they were round and true.

Q.   What causes the wrist-pin in the cross-head and crank-pin to wear the way they do ?

A.   It is simply the motion they have; the crank goes all the way round, forming a circle, and the wrist only vibrates.

Q.   If the cross-head or crank-pin brasses were brass-bound, what should be done ?

A.   They should be chipped and filed.

Q. How do you know when you have taken enough off ?

A. By outside and inside calipers.

Q. How does steam enter the cylinder?

A. In common slide-valve engines it enters through one of the end ports and exhausts back through the same port, when the cavity of the valve has covered it and the exhaust port at the same time. What is a cushion ?

A. Cushion is the resistance on the opposite side of the piston-head, formed by the steam being shut up in the cylinder, as the piston is nearing either dead center.

Q. What is meant by clearance?

A. Clearance is the space between the piston head, cylinder head and valve face at each end of the stroke.

Q. How would you know the amount of clearance there was in that space?

A. By finding the number of cubic inches in a bucket of water, then fill up the space level with the steam port, and see how much water is left in the bucket; the difference is the contents in cubic inches.

Q. Why are gibs, keys and set screws used on both ends of the connecting rod?

A. They are there to take up lost motion.

Q. How would you do that?

A. By loosening up the set screw, and driving down the key; then tighten the set screw to keep the key from raising.

Q. Are there more square inches in one end of the cylinder than in the other?

A. In one sense of the word there are, and in the other there are not, as the piston rod takes up some of the space in one end of the cylinder, therefore there is not the same area in one end as in the other.

Q. What is a governor on an engine for?

A. It is to regulate the steam that passes from the boiler to the steam chest, when the throttle is wide open.

Q, How does it work?

A. It is regulated to allow the engine to run at a certain speed. The governor has a belt from the main shaft to a pulley on the governor. After the engine is running up to the speed it is intended to, it allows only enough steam to enter through the governor valve to keep the same speed; if the engine needs more power it begins to slack up, the governor balls drop, the valve opens and allows more steam to enter; consequently the engine must retain its speed; and if the load is taken off it will start to run away, the governor balls will rise, force the valve shut, and cut off the steam; consequently, the engine must come back to its regular speed.

Q. How does a governor valve look?

A. It is a round valve with grooves; there are different kinds, some have three or four openings, and some only two; the more openings the more sensitive the governor.

Q. Are there other makes of governors?

A. Yes, The Automatic governor on high speed engines, such as are used for running electric light dynamos, they are aronnd the shaft, they work direct on the valve itself.

Q. What is a lubricator?

A. A lubricator is an appliance for holding oil, to be distributed into the valve chest and cylinder, to prevent cutting.

Q. How is it operated?

A. It is operated by steam forcing the oil out of the lubricator into the steam pipe.

Q. Where is the lubricator generally attached.

A. In the steam pipe immediately over the throttle or globe valve, used to start and stop the engine.

Q. State the principal upon which a jet of steam taken from the boiler at boiler pressure can force a stream of water back into the boiler through the injector?

A. It acts upon the principal of a light body moving at a high velocity giving a slower motion to a heavier body effecting an entrance by means of the momentum thus given to it. For instance, steam at the pressure of 80 lbs. to the square inch will escape into the air with a velocity of 1,821 feet per second or 1,241 miles per hour. This rapidly moving jet of steam causes, at first a vacuum in the casing of the injector, which fills with water. The steam then mingles with the water, condenses and imparts its velocity to it. The stream of water is then forced along the pipe and strikes the check valve with a force sufficient to open it and then enters the boiler.

Q. Will the injector work if the water that is supplied is to hot to condense the steam. See p. 95.

LINING AN ENGINE.

Q. How would you line up an engine?

A. By stripping the engine, take off both cylinder heads, if convenient,; then take out the follower-head, piston-rings, bull-ring; disconnect the piston from cross-head: also disconnect the connecting-rod from the cross-head and the crank-pin; then take a slotted stick and place it on one of the studs on the end of cylinder furthest from the crank; then draw a fine seagrass line over the point of stick and through the center of cylinder, and attach it to a stick at the other end of the bed-plate, nailed to the floor or clamped to the bed-plate; then take a thin stick, the length of it being a half inch less than half the diameter of cylinder, and stick a pin in each end of the stick, so they can be forced in or drawn out to suit the adjustment; then center the line at each end of the cylinder at the counter-bore from four sides.

Never center the line in the stuffing box where the piston passes through, but use the inside counter-bore under all circumstances, whether you can remove the back cylinder head or not. Some engine cylinder heads and frames are one; consequently, the head cannot and must not be moved.

Q. If one counter-bore would be out, or larger than the other, what would you do? Would it not throw the bore of the cylinder or the line out?

A. No; center it accordingly; it would not make any difference, only two centering sticks with pins are needed to bring the line central with the bore.

Q. Why do you use the counter-bore?

A. Because the counter-bore is the only true bore the cylinder has that is not worn; consequently, all engineers and machinists must be governed by it. Q. What is a counter-bore?

A. A counter-bore is each end of the cylinder bored from $\frac{1}{16}$ to $\frac{1}{4}$ of an inch larger, from 1 to 4 inches long, according to the size and length of the cylinder.

Q. What is a counter-bore for?

A. To keep the piston from wearing a shoulder in the cylinder at each end.

Q. Why is it that the piston does not wear a shoulder in the cylinder?

A. Because the piston rings just pass over the edge of the regular bore, and by so doing no shoulder can be formed in the cylinder.

Q. How are cylinders bored?

A. They are generally bored on a regular cylinder boring lathe, which has a table that can be raised or lowered to suit. The regular bore is first bored, then the counter-bore, then the two faces for the heads.

Q. How do you square a shaft when you have got the line centrally through the cylinder?

A. Move the crank-pin down to the line and see where the line touches the crank-pin between the two shoulders, then move the pin over to the other dead center, and see how itcomes; if equal, the shaft is square.

Q. If you found it out of square ½ inch, what would you do?

A. Move the out end pillow-block.

Q. Why not move the head-block.

A. Because it would alter the length of the connecting-rod, and be liable to knock out a cylinder-head.

Q. How would you level a shaft?

A. A shaft is leveled by a spirit level, or a plumb-line dropped past close to the line that comes through the cylinder directly in front of the center of shaft; let it drop in a bucket of water to keep the plumb-bob from swaying around; then try the crank pin at both half-strokes (the same principle as in squaring), top and bottom. and see how the crank-pin feels the line; if equal, the shaft is level.

Q. Is there no other way to level a shaft?

A. Yes, by the pulley wheel.

Q. How is it done?

A. Drop a plumb-line down from the ceiling, past the rim's edge of the wheel, directly over the center of the shaft; let the space between the plumb line and rim be one inch; mark the wheel with chalk for a starting and stopping point, and caliper the distance with inside calipers; then turn the wheel and shaft around, and continue calipering until the wheel has made a full revolution; if it calipers the same all the way around, the shaft is level. This principle answers for trueing a wheel as well as leveling a shaft. The former way, by dropping a plumb-line in front of the crank face and feeling the line with the crank-pin at both half-strokes, is the proper way to level a shaft.

Q. If you found the shaft out of level, what would you do?

A. I would have to thin or thicken the brasses, or babbitt the main pillow and out block bearings, whichever the case may be.

Q. How would you know if the center of the shaft is in line with the line through the cylinder or not?

A. It can be found out by placing a two-foot steel square against the crank face, under the line through the cylinder, so that the heel of the square is at the center of the shaft, and see how the square touches the line; if it touches exactly, the shaft is in line; if too hard, the shaft is high; if not at all, the shaft is low.

Q. How would you raise your shaft?

A. There are various ways; by liners, babbitt, heavier or lighter brasses.

Q. If your crank face was oval, and you put a square against it, would that be right ?

A. A spirit level could be placed on a square and bring it level, or drop a plumb-line, and put the end of the square against the crankshaft center, and let it come against the plumb-line. This is a very true way.

Q. Now, after your shaft is in line, square and level, and you still find it out over line $\frac{1}{4}$ inch, what would you do?

A. I would take it off the crank-pin brasses and fill in the other side with a brass ring, or babbitt the side edge of brasses; in some cases the side of the connecting rod has to be chipped to allow it to pass free of the crank-face.

Q. Why would you not take it off the wrist-pin brasses in the cross-head?

A. Because the rod would then be out of the center of cross-head, and have a tendency to bind the piston iu the cylinder and the cross-head in the guides, consequently cutting both.

Q. Would it not make a difference at the other end of the rod?

A. No, the closer the crank-face the better it would be.

Q. Now what would you do?

A. Level and line the guides by putting them in their place, and line them with a pair of calipers, by calipering them at both ends to get them in line with the line through the cylinder, after having found the distance between the side of the cross-head and the center of the cross-head where the piston enters the cross-head. Level by spirit level, first taking spirit level and trying it in the cylinder, if a new one, or on top of the cylinder where it has been planed off when first bored, for they are the only things to go by.

Q. Would you use the valve seat to level by ?

A. No, but alongside of it, where the steam chest rests on.

Q. If you had no spirit level, how would you do it?

A. With a plumb-line, by placing a square

lengthwise on the guides, and try them by bringing the square against the line.

Q. If you had no two-foot square, and could not get any, how would you lay one off?

A. Take a pair of dividers, draw a circle, then find four points on the circle, scribe lines from point to point, which gives a square. This should be done very accurately, or 6,—8 and 10.

Q. Can a plumb-line hang out of true?

A. It can not, provided it hangs clear of everything. If none of these were handy, a straight edge must be placed across the guides at one end, and see if the guides touch the straight edge equally at both edges, then caliper the distance between the line and the straight edge, also at the other end of the guides; if the same, the guides are level lengthwise with the cylinder and line; then level the guides crosswise with a plumb-line and square.

Q. How would you measure the connecting rod of an engine?

A. By finding the striking points.

Q. How would you do that?

A. By shoving the piston and cross-head up

against the cylinder-head, and making a mark
on the guides at one end of the cross-head with
a scriber and center-punch; then move the piston and cross-head back to the other cylinder-
head and make another mark on the guide at
the same end of the cross-head; then measure
from the center of crank-pin to center of shaft;
that gives the half-stroke; double this, gives full
stroke. If half-stroke is 12 inches, the full
stroke is 24 inches; then if the distance between
the two striking points is 25 inches, aad the
stroke 24 inches, the clearance between the cylinder-head and piston-head will be $\frac{1}{2}$ inch when
the piston is at either end of the cylinder.
Then move the cross-head $\frac{1}{2}$ inch back from the
striking point, and bring the crank-pin toward
the same dead center; then take a tram and
measure from the outside center of crank-pin
to the outside center of wrist pin in cross-head,
which will give proper length of connecting-
rod, also the right division of clearance.

Q. What is meant by clearance in the cylinder?

A. It is the unoccupied space between the

piston-head, cylinder-head and valve-face, when the crank-pin is at either dead center.

Q. Does the amount of clearance affect the engine's economy? A. Yes, it does.

Q. How much clearance should there be between the piston and cylinder-head?

A. It depends upon the size; some have from $\frac{1}{4}$ to $\frac{7}{8}$ of an inch.

Q. What is formed in that space or clearance when running? A. A cushion.

Q. What is a cushion?

A. A cushion means the steam that enters the cylinder through the lead the valve has, and the resistance it makes on the piston-head, cylinder-head and valve-face, as the engine is reaching the dead-center.

Q. What is a cushion for?

A. It is to catch the piston and weight of the machinery as it reaches the dead-center, and the lead is to give the engine power at the beginning of the stroke.

Q. How does it act?

A. The same as a spring on the end of a hammer.

Q. If you wished to shorten or lengthen the connecting-rod, how could it be done?

A. By placing tin or sheet iron liners between the brasses and stub-ends of the connecting-rod.

Q. Now, if the key had to be raised, how could this be done?

A. By putting liners between the straps and brasses.

Q. Would that not alter the length of the rod? A. No.

Q. With what instrument would you measure a connecting-rod ?

A. It is called a "tram."

Q. With what is an engine packed in the stuffing-box?

A. Some engineers use hemp, others use black lead packing, and others use lead rings or metalic packing; there are several kinds. Every engineer to his own taste.

VALVE MOTION.

Q. What is an eccentric?

A. An eccentric is a subterfuge for a crank; it is anything out of center.

Q. How would you find the throw or stroke of an eccentric?

A. By measuring the heavy and the light side; the difference between the two is the stroke or throw.

Q. What throw should a common slide valve engine eccentric have?

A, Generally double the width of the entry or steam ports.

Q. If you changed the size of the eccentric would it alter the throw of the valve ?

A. No, it would not, but if you changed the position of the eccentric on the shaft it would.

Q. What is a cam?

A. A cam has no definite meaning; it has 1, 2, 3 or 4 motions; they are used on poppet valve

engines, such as are in use on high pressure
river steamboats.

Q. How would you measure your valve and
eccentric rods?

A. By placing the crank-pin at its dead-center, the center of the eccentric straight or plumb
above the center of the shaft, the rocker-arm
perpendicular, and the valve covering both
ports equally; then take a tram and measure
from the center of the eccentric to the center of
the pin where the eccentric rod hooks on (generally the lower pin) for the eccentric rod, and
from the outside center of the pin where the valve
rod is attached to the furthermost end of the
valve, allowing for two nuts at each end of the
valve, called adjusting and jamb nuts.

Q. How would you plumb an eccentric?

A. By dropping two plumb lines, one at
each side of the shaft, and half the space between the two lines will be where the center of
the eccentric should stand, with heavy side up.

Q. What kind of a tool would you use to
find the exact center?

A. A pair of hermaphrodite calipers, one leg

of which has a sharp point and the other leg has a short foot, so as to feel the line.

Q. What does an eccentric rod consist of ?

A. An eccentric rod consists of a strap, yoke, rod and two nuts; when taking the measure, couple the yoke and strap together, then put a half-inch thick piece of wood between the two straps and find the center of the circle from four sides, with a pair of hermaphrodite calipers, then put the rod in the yoke and adjust it to the proper length by the two nuts; if that will not do, the rod must be shortened or length ened, by cutting out or adding a piece, whichever the case may be. Then take the measure with a tram from the center of the straps to the center of the rod where the rod hooks on lower rocker-arm pin.

Q. How long is the thread on a valve-rod?

A. Long enough to allow two nuts at each end of the valve, and space for adjustment.

Q. Now, if your rocker-arm stood at a quarter, and your eccentric out of plumb, how would you take the measure for the rods?

A. Simply bring them plumb and take

the measure; that is the only right way.

Q. After you have measured the rods, what would you do?

A. They should be put on and the valve set.

Q. What do you move or do first, to set a valve after connections are made?

A. Move the eccentric in the direction the engine is to run, until the valve begins to take steam or lead, then tighten the eccentric temporarily with set screws, then move the crankpin over to the other dead center, and see how much lead it has; if equal the valve is set.

Q. What is meant by the lead of valve?

A. The opening the valve has when the piston is at the beginning of its stroke.

Q. What lead should large engine have?

A. About $\frac{1}{16}$ of an inch. High speed engines must have a quick opening or good lead.

Q. Now if you find the valve laps out $\frac{3}{8}$ of an inch on one end, and the proper lead on the other, what would you do?

A. Divide the difference, by moving the valve one-half it is out, by adjusting the valve-gear. Q. How much?

A. The valve has $\frac{1}{16}$ of an inch lead at one end and laps $\frac{3}{8}$ of an inch at the other end; the valve is out $\frac{7}{16}$ of an inch; then the valve must be adjusted by the nuts one-half it is out, making $\frac{7}{32}$ of an inch. Then throw the crank on the other dead center, move the eccentric whichever way will bring you back to $\frac{1}{16}$ of an inch lead, then tighten temporarily with set screws, throw crank over on the other dead center, and the valve will be set. After valve is set, tighten the eccentric for good.

Q. But if it is not set, what would you do?

A. Go through the same performance until it is set. Some valve-rods have a yoke that slips over the valve, while the adjusting and jam-nuts are between the stuffing box and the rocker-arm pin. When a valve-rod has no nuts, the adjusting must he done at the eccentric rod. To lengthen or shorten the stroke of valve-rod, raise or lower the eccentric-rod pin in the slot, at the bottom of the rocker-arm, whichever way suits the circumstances.

Q. Now, after you have set your valve, keyed everything up properly, and there was a thud or dead sound in the engine or cylinder, what would you do, or where would you look for the trouble?

A. In the exhaust being choked. The steam chest cover must be taken off, then uncouple the valve, turn the valve up sideways and move it until the steam edge has the proper lead with the steam-port, then place a square on the valve-seat of the cylinder, and against the valve-face, to see how the exhaust lead on the opposite steam-port corresponds; if it is choked, then scribe it by allowing a little over double the steam lead.

.Q How is the exhaust made larger?

A. By chipping out the exhaust cavity in the valve, and rubbing file over it to smooth it.

Q. Do you think a little over double the steam-lead would be sufficient for the exhaust?

A. Yes, if not, take out a little more.

Q. Where should the exhaust be?

A. It should be the furthest from the steam-port that is receiving.

Q. What would you do in case your eccentric slipped around on the shaft?

A· Set the valve the same as before.

Q. Is the principle of valve setting the same on all engines?

A. Yes; some engines have two steam and two exhaust valves, but that makes no difference, the principle is the same. (See p. 96).

Q. How would you find the dead center of an engine?

A. By placing a spirit level on the strap that goes around the brasses that connect the crank-pin to the connecting-rod, and when it is level the crank is at a dead center. If the engine is not level, then use an adjnstible level.

Q. What other way could you find the dead center of an engine?

A. By moving the engine toward the dead center until the cross-head stopped moving; then put a center punch mark in the floor, and one on the fly-wheel, after having marked it with a tram; then move the crank over the center until the cross-head begin to move, then put another mark; the middle between the two marks is the exact dead center; then bring the middle mark to the point of the tram; this is done with a small tram with one straight point and a short foot.

Q. If the engine had to be run in the opposite direction to which it had been running, how could it be done?

A. It could be done by placing the crank-pin on the dead center, removing the steam-chest cover, and turning the eccentric over on the shaft in the opposite direction, until the valve has the proper lead at the opposite port, then try the engine from dead center to dead center, to equalize the lead at both ends of the valve; then the engine will run in the opposite direction.

Q. Does a crank-pin and piston travel the same distance?

A. No, a crank-pin travels $1\frac{14116}{10000}$ times further than the piston each revolution, or $0.\frac{5707}{10000}$ times further each stroke. For example, take an engine with a 12-inch stroke, the piston travels 24 inches and the crank pin $37\frac{6992}{10000}$ inches each revolution, or the piston travels 12 inches each stroke and the crank-pin 18.8496 per single stroke of piston. To do this, multiply the single stroke by one-half of 3.1416, which is 1.5708, and the answer will be the distance the crank-pin travels further than the piston per single stroke. This rule answers for all engines. Another fact not generally known by many men is that a crank of an engine, at two certain points, travels a long distance while the motion of the cross-head is hardly noticed. When the center of the crank-shaft and crank-pin are in a line with the piston-rod, no steam pressure applied to either side of the piston can set the engine in motion; this is called the dead center.

Q. Is the piston-head in the center of the cylinder when the centers of the crank-pin and crank-shaft are plumb, or in right angles with the cylinder?

A. No, under no circumstances.

Q. What is a revolution?

A. It means the crank has turned once around, or made a circle.

Q. How many strokes has a revolution?

A. Two to each revolution.

Q. If an engine has 24 inches stroke, and

makes 65 revolutions per minute, how many feet does it travel in a minute?

A. Twenty-four inches multiplied by 2 equals 48 inches, this multiplied by 65 revolutions equals 3120 inches, which divided by 12 equals 260 feet per minute.

Q. If you were asked the horse power of any sized engine, could you tell it? A. Yes.

Q. Well, how would you go about it, and what is a horse power?

A. A horse power is 33,000 pounds raised 1 foot high in 1 minute, or 150 pounds raised 220 feet high in 1 minute. To find the horse power of any engine, first find the area of the piston-head face, then multiply the answer by the average pounds pressure per square inch in cylinder, then multiply by the number of feet traveled in 1 minute, and divide by 33,000.

EXAMPLE:

Cylinder 12 x 24 in.   12   diam. of cylinder.
65 revolutions,   12

Average pressure 40 ℔s.   144   sq. of diameter.
.7854

Generally allow about ½ the boiler pressure in figuring H. P.

113.0976 area of p. h. face.
40 average pressure in
——   the cylinder.
4523.9040
260 No. ft. trav. by p.

33000)1176215.0400(35.6428 I. H. P.

THE INDICATOR.

The steam engine indicator is an instrument for showing the pressure of steam in the cylinder at all points of the stroke, or for producing actual diagrams. The indicator consists of a small cylinder accurately bored out, and fitted with a piston, capable of working in the (indicator) cylinder with little or no friction, and yet be practically steam-tight. The piston has an area of just $\frac{1}{2}$ of a square inch, and its motion in the cylinder is $\frac{25}{32}$ of an inch.

The piston-rod is connected to a pair of light levers, so linked together that a pencil carried at the center of the link moves in nearly a straight line through a maximum distance of $3\frac{1}{8}$ inches. A spiral spring placed in the cylinder above the piston, and of a strength proportioned to the steam pressure, resists the motion of the piston; and the elasticity of this spring is such that each pound of pressure on the piston causes the pencil to move a certain fraction-

;al part of an inch. The pencil in this case is made of a piece of pointed brass wire, which retains its sharpness for a considerable time, and yet makes a well-defined line upon the prepared paper generally used with the indicator.

The paper is wound around the drum, which has a diameter of 2 inches, and is capable of a semi·rotary motion upon its axis to such an extent that the extreme length of diagram may be $5\frac{1}{4}$ inches. Motion is given to the drum in one direction, during the forward stroke of the engine, by means of a cord connected indirectly to the cross-head of the engine, and the drum is brought back again during the return stroke of the engine by the action of a coiled spring at its base.

The conical stem of the instrument permits it to be turned around and fixed in any desired position, and the guide-pulleys attached to the instrument under the paper drum may also be moved around so as to bring the cord upon the drum-pulley from any convenient direction.

The upper side of the piston is open to the atmosphere; the lower side may, by means of a stop-cock, be put into communication either with the atmosphere or with the engine cylinder.

When both sides of the piston are pressed upon by the atmosphere, the pencil, on being brought into contact with the moving paper, describes the atmospheric line. When the lower side of the piston is in communication with the engine cylinder, the position of the pencil is determined by the pressure of the steam existing in the cylinder;. and on the pencil being pressed against the paper during a complete double stroke of the engine, the entire indicator diagram is described.

In order that the diagram shall be correct, the motion of the drum and paper shall coincide exactly with that of the engine piston; second, that the position of the pencil shall precisely indicate the pressure of steam in the cylinder; third, that the pendulum must be from $1\frac{1}{2}$ to 3 times as long as the stroke of the engine piston; fourth, that the pendulum must be plumb when the piston is at half-stroke; fifth, that the cord around the drum must be attached to the pendulum at right angles, or square with the indicator; sixth, the pendulum must be attached with an inch wooden pin to the ceiling or floor at one end, the other end to the cross-head by means of a screw bolt in the wrist-pin and a slot in the pendulum; seventh, that the two holes tapped in the cylinder are directly opposite the steam ports, and centrally between the piston-head and cylinder head, when the engine is at the dead center, or, in other words, in the center of clearance; eighth, that the piping should be as short as possible, and $\frac{1}{2}$ inch pipe if not over 1 foot long. If longer the pipe should be larger close to the cylinder, and covered so as not to allow too much condensation, as it affects the diagram. The best way to take a diagram is to tap a hole in each cylinder-head and take each end separately. The cord must be attached to the pendulum, so the paper drum will move in proportion to the piston.

An indicator shows the highest and the lowest pressure reached, also the cut-off and lead. If there is a great difference, say more than 5 pounds, between the boiler pressure and the initial pressure upon the piston, the connecting pipes may be taken as being too small, too abrupt, or the steam ports too contracted. The full pressure of steam should come upon the piston at the very beginning of its stroke. Should the admission corner be rounded, the valve is wanting in "lead," or, in other words, the port for the admission of steam is uncovered too late in the stroke.

The steam line should be parallel or straight with the atmospheric line up to the point of cut-off, or nearly so. Should it (the steam line) fall as the piston advances, the opening for the admission of steam is insufficient, and the steam is "wire-drawn."

The point of cut-off should be sharp and well defined; should it be otherwise, the valve does not close quick enough. The bevel line leading from the cut-off line to the end of the stroke is called the expansion line.

Q. Which is the standard indicator?

A. The Tabor's improved.

Q. Are there any other makes?   A. Yes; Richard's, McNought's, Thompson's and others.

RULES.

RULE for telling the power of a diagram: Set down the length of the spaces formed by the vertical lines from the base in measurements of a scale accompanying the indicator, and on which a tenth of an inch usually represents a pound of pressure; add up the total length of all the spaces, which will give the main length, or the main pressure upon the piston in pounds per square inch; to do this, lay a card taken by the indicator off in ten parts, by drawing lines from top to bottom. Find out what the scale is; suppose it is 60, the number of ordinates 10, and that the sum of their length is 6 inches; so 6 and 10 ordinates $= \frac{6}{10}$ or 6 x 60 = 36.0. Answer, 36 pounds pressure upon the piston.

RULE for finding and deducting friction: Multiply N. H. P. by .13 and subtract the answer from N. H. P., which gives I. H. P.

Q. What is N. H. P.?

A. It is nominal horse power.

Q. What is I. H. P.?

A. It is indicated horse power.

Q. What is meant by cutting off steam at 6 inches?

A. It means that the valve closes and cuts off the live steam from the boiler at 6 inches of the piston's travel; then the engine gets its power, from the time the valve closes or cuts off until the exhaust opens, by the expansion of the steam closed up in the cylinder.

Standard multiplers, with examples :

1. For the Area of a Circle.	Multiply sq. of diam, by	.7854
2 For Circumference of a Circle,	Multiply diameter by	8.1416
3. For Diameter of a Circle,	Multiply the circum. by	.31831
4. For the Surface of a Ball,	Multiply sq of diam. by	3.1416
5. For the Cubic Inches in a Ball,	Multiply Cube of dia. by	.5236

1. RULE for finding the area of any circle. Always multiply the diameter by itself, then by .7854, then cut off 4 decimals to the right.

2. RULE for finding the circumference of anything round. Multiply the diameter by 3.1416, and cut off 4 decimals.

3. RULE to find diameter of circle. Multiply circumference by .31831.

EXAMPLE: The circumference 9.4248 x .31831 = 3.000008088 = 3 inches diameter.

4. RULE to find the surface of a sphere, globe or ball.

EXAMPLE: 9 inches diameter x 9 = 81 x 3.1416 = 254.4696.

5. RULE to find the cubic inches in a ball. Multiply cube of the diameter by .5236; the answer equals its solid contents.

EXAMPLE: Ball 3 inches in diameter; 3 x 3 = 9 9 x 3 = 27 x .5236 = $14\frac{1372}{10000}$ solid contents.

RULE to find pressure on the crown sheet of a hanging fire-dox boiler. Multiply the width by the length in inches, then multiply by steam gauge pressure and devide by 2.

EXAMPLE:

Crown sheet 46 x 33 in.   46
Pressure 85 lb.   33
                               ————
Iron ½ in.   1518
                                  85
                               ————
If iron is ¼ in. div. by 4.   2)129030
                               ————
If iron is ⅜ in. div. by 2.66 2000)64515 lbs. press'e.
                               ————
                             32.$\frac{257}{}$ tons "

RULE to find how much water a boiler will contain. For 2-flue boiler, ⅔ full of water, find ⅔ of the area of the boiler in inches inside; multiply by length in inches; then find area of flues, thickness of iron added; then mnltiply by 2, if two flues; multiply by length in inches, subtract area of flues from ⅔ contents, and divide by 231 (number of cubic inches in a standard gallon); the answer will be the number of U. S. gallons.

EXAMPLE:

Boiler 48 inches.	48
Two flues, 16 in. each.	48
Length 20 feet.	2304
16	.7854

16	3)1809.5616 area of Boiler.
256	603.1872 One-third of area.
.7854	2
·201.0624	1206.3744 Two-thirds of area
2	240 Length in inches.
402.1248	289529.8560
240	96509.9520 Sub. Area of Flues.
96509.9520	231)193019.9040
	835.5940 No. of Gallons.

RULE to find the amount of water required, when the average pounds of coal used per hour is known. Divide the coal by 7.5; the answer will be cubic feet; then multiply by 7.5, and that gives the number of U. S. standard gallons.

EXAMPLE:

117 lbs. of coal used per hour, 7.5]117.0

$$\frac{15\text{-}6}{7.5}$$

117-0= 117 gals.

Q. How many cubic feet in 1 lb. of air?

A. $13\frac{817}{1000}$ cubic feet.

Q. How much air does it take to consume 1 pound of coal?

A. It takes 18 pounds, or $248\frac{706}{1000}$ cubic feet.

Q. How would you tell the amount of water any tank contained?

A. If the tank was large at the bottom and narrow at the top, lay the tank off in 10 parts from top to bottom, then take the diameter $\frac{4}{10}$ from the large end of the tank, square it, then multiply by .7854; that gives the area; then multiply quotient by full depth of tank and divide by 1728, which gives the number of cubic feet; multiply answer by 7.5, and the number of U. S. gallons will be given. The example must be done in inches; 1728 is the number of inches in a cubic foot, and 7.5 is the number of gallons in a cubic foot.

EXAMPLE:

Tank 2 feet diam.   24 inches diameter.
Tank 3 feet deep,   24   "   "

$$576$$
$$.7854$$
_____

452.3904 area in inches.
36 inches deep.
_____

1728)16286.0544
_____

9.4218 cubic feet.
7.5   No. gals. in a cub. ft.
_____

70.86000 U. S. gals. in tank.

RULE how to mark engineer's tools. Warm the tool and allow a thin coat of beeswax to cover the place to be marked; after the beeswax is cold, take a dull scriber and do the marking; then apply some nitric acid, after a few moments wash off the acid with water, then heat the tool to melt the beeswax, and you will find well defined marks.

RULE for chimneys. Chimneys should be round inside, instead of square, to insure a good draft. The opening should be one-fifth larger than the area of the flues or tubes combined; if less, the draft will not be free. The opening from the bottom should increase in size to the top, and be smooth inside.

Rule for making good babbitt metal, for high and low speed, in parts.

HIGH SPEED.		COMMON.		MEDIUM.	
Martin's Nickel.	10	Copper..........	12	Copper..........	60
Copper..........	16	Antimony......	4	Antimony......	25
Antimony.......	4	Tin	84	Tin......•.......	15
Tin	70				
	100		100		100

Rule for babbitting a box. Nearly every engineer has his own way; but the best and quickest way is to chip out all the old babbitt in the cap and box, then put the journal or shaft that is to run in the box in its place; put enough liners in between the shaft or journal and edge of box until level, square and in line; put thick putty around the shaft and against the box, so the babbitt can not run out; then heat the babbitt until it runs free, and pour accordingly; the cap is then bolted in its place upon $\frac{1}{16}$ inch thick liner, and putty placed as before; then pour metal through the oil holes, which will have to be drilled out afterwards.

RULE to determine the capacity of any size pump, single or double action. Multiply the area of the water piston-head face or plunger in inches, by its stroke in inches, which gives the number of cubic inches per single stroke; the answer divided by 231 (the cubic inches in a gallon) will give the number of standard gallons per single stroke. But remember, all pumps throw less water than their capacity, which depends upon the condition and quality of the pump. This loss arises from the rise and fall of the valves; from a bad fit or leakage, and in some cases from there being too much space between the valves, piston or plunger. The higher the valves have to rise to give the proper opening, the less work the pump will perform.

Q. Will a boiler 60 inches in diameter, $\frac{3}{8}$ inch iron, stand as much pressure as a boiler 48 inch diameter, $\frac{3}{8}$ inch iron? A. No.

Q. Why?

A. Because the pressure in the large boiler has more surface, and will not allow it. It is the same as a long bar and a short bar of the same thickness; it takes less strain to break the long one than the short one.

RULE for finding safe working pressure of steam boilers. Always use .56 for single riveted and .70 for double riveted side seams. A radius means ½ the diameter and ⅕ of tensile strength is safe load. U. S. standard is ⅙.

Multiply the thickness of iron by single or double rivets, then multiply by the safe load, divide by internal radius, and the answer will be the safe working pressure.

EXAMPLE:

Diam. 42 in.   .1875 thickness of iron.

Iron ₃⁄₁₆ in.   .70 double riveted.

Double riveted   .131250

50,000 lbs. tensile str'th. 10000   2)42

20.8125)13125000.00   21 outside radius.

Safe working pressure,   63.06   .1875

                            5   20.8125 inside radius.

Bursting pressure,   315.30

RULE to find aggregate strain caused by the pressure of steam on the shells of boilers. Multiply the circumference in inches by the length in inches; multiply this answer by the pressure in pounds. The result will be the pressure on the shell of boiler, and divide by 2000, which gives the tons.

EXAMPLE:

Diam. of boiler 48 inches, circumference 150.7968, length 20 feet, or 240 inches, pressure of steam 120 lbs. 150.7968 x 240 x 120 = 4342947.8400 lbs., divided by 2000 = 2171½ tons strain.

RULE to find the number of feet of 1 inch pipe required to heat any size room with steam. For direct radiation 1 lineal foot (straight foot) to 25 cubic feet of space. For indirect radiation, 1 lineal foot to 15 cubic feet of space. Note, all pipe is measured inside for size.

EXAMPLE:

Room 18 x 18 x 18 to be heated with 1 inch pipe. Direct radiation. All calculating must be done in inches, and divided by 1728 to find the cubic feet.

$$216$$
$$-\ 216$$
$$\overline{46656}$$
$$216$$
$$\overline{1728)10077696}\text{ cubic inches.}$$
$$25)5832\text{ cubic feet.}$$

Lineal 233$\frac{7}{25}$ feet of 1 inch pipe.

One cubic foot of boiler is required for every 1500 cubic feet of space to be warmed. One horse power of boiler is enough for 40,000 cubic feet of space.

RULE to find the horse power of a boiler. Always find the number of square inches and divide by 144, which gives the square feet of heating surface, and divide by 15 square feet, whcih is an average allowance for one horse power of a boiler; divide the H. P. by 2, you will have the proper grate surface, and allow ½ square inch of safety valve to each square foot of grate surface. Generally, from ½ to ¾ of a square foot of grate surface is allowed to each horse power of a boiler.

Q. How do you find the horse power of a boiler?

A. Find the number of square feet of heating surface and divide by 15; 15 square feet of heating surface is the general allowance for a H. P. of a boiler. (See following example.)

EXAMPLE:

Boiler 48 in. x 25 ft.　　First find circum. of boiler.
Two 16 in. flues.　　　　16 in. diam. of 1 flue.
　48 diam. of shell.　　　　3.1416
　　3.1416　　　　　　　 ─────────
─────────　　　　　　　 50.2656 circ. of 1 flue.
3)150.7968　　　　　　　　　300 length of flue.
─────────　　　　　　　 ─────────
　50.2656 one-third circum. 15079.6800 in inches.
　　2　　　　　　　　　　　 2
─────────　　　　　　　 ─────────
100.5312 two-thirds "　　30159.3600 heat. sur. 2 fl.
　300 length or boiler in inches.
─────────
30159.3600 No. sq. in. heat. surf.　16 in diam. of 1 flue.
　　　　in the shell.　　　　16
　48　　　·　　　　　　　　 ───
　48　　　　　　　　　　　 256
─────　　　　　　　　　 .7854
2304　　　　　　　　　　 ─────────
　.7854　　　　　　　　 201.0624 area 1 flue.
─────────　　　　　　　　 2
3)1809.5616 area of 1 head.　402.1248 area 2 flues,
─────────
　603.1872 one-third area of 1 hd.　2
　　　　　　　　　　　　 ─────────
　　2　　　　　　　　　 804.2496 both ends.
─────────
1206.3744 two-thirds area of 1 hd.
　　2
─────────
2412.7488 two-thirds area of both heads.
No. sq. in. heat. surf. in shell, 30159.3600
　"　　　"　　"　　flues, 30159.3600
Two-thirds area both heads,　 2412.7488
　　　　　　　　　　　 ─────────
　Total,　　　　　　　 62731.4688
Subtract area of flues,　　 804.2496
　　　　　　　　　　　 ─────────
　This boiler is 28 h. p. An 144)61927.2192
engine uses about ½ of boil-　15(430.　 sq. ft. h. s.
er's h. p.. making this boil-　2(28.　　h. p.
er sufficiently large enough　 2(14.　　grate surf.
to supply engine of 56 h. p.　　7.　　area s'fty v.

Number sq. feet of heating surface allowed for tubular boilers are 12 sq. feet. Flue boilers 15 sq. feet. Cylinder boilers 7 sq. feet.

RULE to find the horse power generated in any kind of boiler when running. First, notice how long it will take to evaporate 1 inch of water in the glass gauge, divide this into 60, which gives the number of inches evaporated in one hour; second, multiply the average diameter where evaporation took place by the length of the boiler in inches; this multiplied by the number of inches evaporated, and the answer divided by 1728 gives the cubic feet of water evaporated in one hour.

As a rule, 1 cubic foot of water evaporated is generally allowed for 1 horse power; also the capacity of a pump or injector for any boiler should deliver 1 cubic foot of water each horse power per hour, and an engine uses one-third of a cubic foot of water per horse power.

EXAMPLE:

Length of boiler 216 inches.   216
Average diam.   40 inches.   40
One inch evaporated in   15(60   8640
    15 minutes.   4   4

1728)34560(20 h. p.

Weight of Square Superficial Foot of Boiler Plate when Thickness is Known.

Thickness.		Weight.	Thickness.		Weight.
Inches.	Dec.	lbs.	Inches.	Dec.	lbs.
$\frac{1}{32}$ = .03125		1.25	$\frac{5}{16}$ = .3125		12.58
$\frac{1}{16}$ = .0625		2.519	$\frac{3}{8}$ = .375		15.10
$\frac{3}{32}$ = .0937		3.788	$\frac{7}{16}$ = .4375		17.65
$\frac{1}{8}$ = .125		5.054	$\frac{1}{2}$ = .5		20.20
$\frac{5}{32}$ = .1562		6.305	$\frac{9}{16}$ = .5625		22.76
$\frac{3}{16}$ = .1875		7.578	$\frac{5}{8}$ = .625		25.16
$\frac{7}{32}$ = .2187		8.19	$\frac{3}{4}$ = .75		30.20
$\frac{1}{4}$ = .25		10.09	$\frac{7}{8}$ = .875		35.30
$\frac{9}{32}$ = .2812		11.38	1 = .1		40.40

Q. Explain how the above fractional parts of whole numbers are made to read as decimals —take $\frac{3}{16}$ of an inch for an example?

A. To do this take 100 as a whole number; divide 16 into 100 = 6$\frac{1}{4}$, reads .625 = $\frac{1}{16}$ of 100. $\frac{3}{16}$ would read, 3 × .625 = .1875. This principle answers for all the rest.

RULE for safety valves. To find the distance ball should be placed on lever, when the weight is known, or the distance is known and weight is not known. Multiply the pressure required by area of valve, multiply the answer by the fulcrum; substract the weight of the lever, valve and stem, and divide by the weight of ball for distance, or divide by distance for weight of ball with the same example as follows:

EXAMPLE:

Weight of ball,   60 lbs.   100 lbs. pressure.
Pressure,   100 "   3 area of valve.
Wt. of L. V. & steam, 30 " $\overline{300}$
Fulcrum,   4 inch,   4 fulcrum.
Area of valve,   3 " $\overline{1200}$
                  30 wt. of L. V. & st.
          60)$\overline{1170}$
                19½ inch ball should
                be hung on lever.

The mean effective weight of valve, lever and stem is found by connecting the lever at fulcrum, tie the valve-stem to lever with a string, attach a spring scale to lever immedi-ately over valve, and raise until the valve is clear of its seat, which will give the mean effective weight of lever, valve and stem.

RULE for figuring the safety valve and to know the pressure when the area of valve, the weight of lever, valve and stem. the distance fulcrum is from valve, and weight of ball is known.

Divide fulcrum into length of lever, multiply

answer by weight of ball, add weight of lever, valve and stem, and divide by area of valve. Answer will be steam pressure.

Weight of ball,	50 lbs. 2.25	4)20
Wt. of L. V. and stem,	30 lbs. 2.25	5
Fulcrum,	4 in. 5.0625	50
Diam. of valve,	2¼ in. .7854	250
Length of lever,	20 in. 3.97608750 area.	30

Add as many ciphers to the divi-   3.9)280.0

dend as there are decimels in the di- lbs.press. 71.$\frac{31}{39}$ visor, and divide as whole numbers.

To measure or mark off the lever, you measure the fulcrum and make notches the same distance as fulcrum; if fulcrum is 4 inches, each notch must be 4 inches apart.

Q. What is meant by a fulcrum?

A. The distance valve stem is from where the lever is connected.

Rules for Machinists.

Rule to Gear a Lathe for Screw-Cut-
ting.—Every screw cutting lathe contains a
long screw called the lead screw, which feeds
the carriage of the lathe while cutting screws;
upon the end of this screw is placed a gear to
which is transmitted motion from another gear
placed on the end of the spindle; these gears
each contain a different number of teeth, for
the purpose of cutting different threads, and
the threads are cut a certain number to the
inch, varying from one to fifty. Therefore, to
find the proper gears to cut a certain number of
threads to the inch, you will first multiply the
number of threads you desire to cut to the inch
by any small number, 4 for instance, and this
will give you the proper gear to put on the lead
screw. Then with the same number, 4, multiply the number of threads to the inch in the
lead screw, and this will give you the proper

gear to put on the spindle. For example, if you want to cut 12 to the inch, multiply 12 by 4, and it will give you 48. Put this gear on the lead screw, then with the same number 4, multiply the number of threads to the inch in the lead screw. If it is 5, for instance, it will give you 20; put this on the spindle and your lathe is geared. If the lead screw is 4, 5, 6, 7 or 8, the same rule holds good. Always multiply the number of threads to be cut first.

Some—indeed, most small lathes—are now made with a stud geared into the spindle, which stud only runs half as fast as the spindle, and in finding the gears for these lathes you will first multiply the number of threads to be cut, as before, and then multiply the number of threads on the lead screw as double the number it is. For instance, if you want to cut 10 to the inch, multiply by 4, and you get 40; put this on the lead screw, then, if your lead screw is 5 to the inch, you call it 10, and multiply by 4, and it will give you 40. Put this on your stud and your lathe is geared, ready for cutting.

Rule for Cutting a Screw in an Engine

LATHE.—In cutting V-thread screws, it is only necessary for you to practice operating the shipper and slide screw-handle of your lathe before cutting. After having done this until you get the motions, you may set the point of the tool as high as the center, and if you keep the tool sharp you will find no difficulty in cutting screws. You must, however, cut very light chips, mere scrapings in finishing, and must take it out of the lathe often, and look at it from both sides very carefully, to see that the threads do not lean like fish scales. After cutting, polish with a stick and some emery and oil.

RULE FOR CUTTING SQUARE THREAD SCREWS.—In cutting square thread screws, it is always necessary to get the depth required with a tool somewhat thinner than one-half the pitch of the thread, after doing this make another tool exactly the pitch of the thread and use it to finish with cutting a slight chip on each side of the groove. After doing this, polish with a pine stick and some emery. Square threads for strength should be cut one-half the depth of their pitch, while square threads for wear may—and should be—cut three-fourths the depth of their pitch.

RULE FOR MONGREL THREADS.—Mongrel, or half V half square threads, are usually made for great wear, and should be cut the depth of their pitch, and for extraordinary wear they may be cut $1\frac{1}{2}$ the depth of the pitch. The point and the bottom of the grooves should be in width $\frac{1}{4}$ the depth of their pitch. What is meant here by the point of the thread is the outside surface, and the bottom of the groove is the groove between the threads. In cutting, these threads, it is proper to use a tool the shape of the thread, and in thickness about $\frac{1}{5}$ less than the thread is when finished. As it is impossible to cut the whole surface, at once, you will cut it in depth about $\frac{1}{16}$ at a time then a chip off the sides of the thread, and continue in this way alternately till you have arrived at the depth required. Make a gauge of the size required between the threads and finish by scraping with water. It is usually best to leave such screws as these a little large until after they are cut, and then turn off a light chip, to size them; this leaves them true and nice.

RULE TO TEMPER TOOLS USED DAILY, SUCH AS CHISELS, TAPS, DIES, REAMERS, TWIST DRILLS, COMMON FLAT DRILLS, AND LATHE TOOLS.—To temper flat, cape or side chisels, and common flat drills, put the tool to be tempered in the fire and heat slowly to a cherry red color, about 4 inches from the point. Then take it out and put it in the water, point first, about three or four inches, then draw it back quick about an inch from the point, and leave it so until the water will barely dry on the chisel, then take it out, polish it with a piece of sand stone, and let the heat that is left in the body of the tool force its way toward the point; it will be noticed immediately in the change of color. The color of temper for chisels to cut cast iron should be a dark straw, turning to a blue. The temper of chisels to cut wrought iron or steel should be plunged into water after the dark straw color has disappeared and the blue begins to show itself, and left in the water to cool off. In some cases, where the tool is too cold and the temper will not draw, put the tool in and out of the fire often, until the temper shows itself, then cool immediately. If the temper gets to the point of tool before it is polished, it will have to be heated over again. The above rule answers for lathe, plainer and shaper tools as well.

Taps, dies, reamers and twist drills should be tempered in oil. After being heated to a cherry red all over equally, drop the tool into a bucket of oil (plumb) and leave it there until cold; then take it out and brighten it with emery cloth; be careful not to drop it, because it is brittle and liable to break. To draw the temper of taps, reamers and twist drills, heat a heavy ring red hot and enter the tool centrally in the ring, so the heat will be equal from all sides. The hole in the ring should be about three times the diameter of the tool. An old pulley hub would be about right. The color for reamers, taps and twist drills should be dark straw, turning to a blue near the shank; where the color is changing too fast, drop a little water on it; after the right color is obtained, cool off in water. To draw the temper in dies after being cooled in oil, set them (the threads up) on a piece of red-hot iron and draw temper the same color as taps.

For tempering a spring, heat it cherry red and put it in oil; after it is cool, take it out and hold it over the fire until the oil burns off; then put the spring in the oil again, then in the fire; do this three times; after the last time, plunge it into water and cool off.

A. No. Q Why. A. Because steam is highly elastic and bulky, and, of itself, would have no effect in driving the hot water in any particular direction. But when steam is moving at a high velocity and is condensed these particles of water have the power of driving the main body before it into the boiler.

The principal is easily explained, for instance. If a block of wood is laid upon the water it will float, but if it is thrown violently downward it will at first go below the surface. Then if there were something there to catch it and hold it, we would have a state of affairs similar to the injector, where the water enters the boiler by its own momentum and is held there by the check valve.

Adjustment and Setting of Corliss Engine Valves.

It often happens that engineers, under whose control Corliss engines are, placed, are not practically acquainted with the operation of the Corliss valve gear, and are at a loss what to do should the gear need adjustment. By carefully observing the following questions and answers, the desired information will be found.

Q. Into how many classes are the different types of Corliss valve gear divided?

A. Into two general classes.

Q. Which are they?

A. To the first class belong the crab-claw gear, originnally used by George. H. Corliss, and later by Harris and other prominent Corliss engine builders.

To the second class belong the half-moon valve gear, as used on the Reynolds Corliss engine built several years since, and followed in some recent designs of Corliss engines.

Q. Which is the more favorable and widely known type now in general use?

A. The half moon type.

Q, Why so?

A. Because the old style crab-claw steam valve opens toward the centre of the cylinder, which obstructs the supply passage and forces the steam to pass over and around the valves. This fault is overcome in the half moon type, as the steam valve opens away from the center of the cylinder, thus leaving a clear and direct passage for the steam into the cylinder.

Q. Do the two different styles make any difference into the opening of the exhaust valves?

A. No. The difference in the two classes is simply in the direction of movement of steam valves; the exhaust valves open the same in either class, viz.: away from the center of the cylinder.

Q. What name has the Corliss valve gear?

A. It is called a detachable valve gear.

Q. Why is it called detachable?

A. Because the steam valves open positively

at the proper time by the direct action of the working parts of the engine, and continue to open until the connection with the working parts of the engine are broken by detaching or tripping the hook, by action of the cut-off cams.

Q. How are the steam valves closed?

A. When the steam valves are detached they are closed by the action of springs, weights, or more generally vacuum dash pots, thus cutting off the supply of steam.

Q. How is the detachment or tripping determined?

A. The time in the stroke at which the tripping takes place is known by the position of the cut-off cams, which are moved and controlled by the governor.

Q. Does the cut-off cams trip the hook always at the same point?

A No. The cut-off is determined by the requirements of the load on the engine.

Q. By what name is this cut-off known?

A. The automatic cut-off.

Q. How is the theory of the Corliss valve motion easily understood?

A. The theory is easily understood by considering the four valves as the four parts (or edges) of a common slide valve.

Q. Why are the four valves of the Corliss engine considered as the four parts (or edges) of the common slide valve?

A. The working edges of the two steam valves answering as the two steam edges of the slide valve, and the working edges of the two exhaust valves as the exhaust edges of the slide valve.

Q. The Corliss having four valves, and the common slide valve only one, does it not make any difference in setting?

A. As far as the setting the principle is the same; the only difference is in the adjustment.

Q. Why does the adjustment make a difference?

A. The four working edges of the common slide valve are in one solid valve, so that any change or adjustment of one of the edges interferes with the other three. If one edge is to be changed in reference to the others, it must be done by altering the valve itself. The Corliss valves, on the other hand, are adjustable, each by itself, and any one of the valves may be changed without disturbing the other three.

Q. Can the adjustment be made while running?

A. When the engineer is familiar with his engine and knows what changes are necessary, the adjustment may be, and is frequently, made without stopping the engine.

Q. How many edges has a slide valve?

A. Four—two steam and two exhaust .

Q. Has the Corliss valves the same number of edges as the common slide valves?

A. No. Each Corliss valve represents an edge of the common slide valve, viz.: two steam edges, two steam valves, two exhaust edges, two exhaust valves.

Q. How are the valves connected to the eccentric and worked on Corliss engines?

A. With the wrist-plate, carrier arm, rocker arm and reach rod.

Q. Is the wrist-plate good for any other purpose?

A. Yes. It modifies the speed of travel at

different parts of the stroke, in relation to each other, and gives a quick and constantly increasing speed when opening the steam valves, and a quick opening and closing of the exhaust valves.

Q. When do the steam and exhaust valves travel slowest? A. When they are closed.

Q. Can the valves of Corliss engines be adjusted when the reach rod is unhooked from the wrist-plate, so the valves may be properly set, independent of the position of the crank?
A. Yes.

Q. Are the Corliss valves easily set?

A. If the engineer has any knowledge, as he should have, of the ordinary slide valve, and of the effect of "lap and lead" as applied to its working, and will consider the Corliss valve gear in the light of this knowledge, he will soon master the seeming difficulties in his way and find the Corliss gear to be the simplest, most perfect and most easily adjusted of all valve motions.

Q. How would you go about setting the Corliss valves?

A. Begin by taking off the back caps or back heads of all four valve chambers. Guide lines will be found on the ends of the valves and on the ends of the chambers, as follows: On the steam valves, coinciding with the working edges of the valves; on the steam valve chambers, coinciding with the working edges of the steam ports. On the exhaust valves and ports, guide lines are also scribed to set them by. The wrist-plate is centrally between the four valve chambers, on the valve gear side of the cylinder. A well defined line will be found on the stand which is bolted to the cylinder, and three lines on the hub of the wrist-plate, which, when they coincide with the line on the stand, show the central position of the wrist-plate and the extremes of its throw or travel. To adjust the valves, first unhook the reach rod connecting wrist-plate with rocker arm and place and hold the wrist-plate in its central position. The connecting rods between steam and exhaust valve arms and wrist-plate are made with right and left hand screw threads on their opposite ends, and provided with jamb nuts, so that ϊy slacking the jaml. nuts and turning the rod they can be lengthened or shortened as desired. By means of this adjustment. set the steam valves so that they will have $\frac{1}{4}$ inch lap for 10 inch diameter of cylinder, and $\frac{1}{2}$ inch lap for 32 inch diameter of cylinder, and for intermediate diameters in proportion.

For the exhaust, set them with 1-16 inch lap for 10 inch bore, and $\frac{1}{8}$ inch lap for 32 inch bore on non-condensing engines and nearly double this amount on condensing engines, for good results. Lap on the steam and exhaust valves will be shown by the lines on the valves being nearer the center of the cylinder than the lines on the valve chambers. Having made this adjustment of the valves, the rods connecting the steam valve arms with the dash pots should be adjusted by turning the wrist plate to its extremes of travel and adjusting the rod so that when it is down as far as it will go, the sq. steel block on the valve arm will just clear the shoulder on the hook. If the rod is left too long, the steam valve stem will be likely to be either bent or broken; if too short, the hook will not engage, and consequently the valve will not open. Having adjusted the valves as stated, hook the engine in and, with the eccentric loose on the shaft, turn it over and adjust the eccentric rod so that the wrist-plate will have the correct extremes of travel, as indicated by the lines on back of hub of wrist-plate. Then place the crank on either dead center and turn the eccentric in the direction in which the engine is to run to show an opening at the steam valve of from 1-32 to $\frac{1}{8}$ inch, depending upon the speed the engine is to run. This opening will be shown by the line on the valve being nearer the end of the cylinder than the line on the valve chamber. This opening gives the "lead" or port opening when the engine is on the dead center. The faster the engine is to run the more lead it requires, as a general rule. Having turned the eccentric so as to secure the desired amount of lead, tighten it securely, by means of the set screw, and turn the engine over to the other center, and note if the other steam valve has the same lead. If not, adjust by lengthening or shortening the connecting rod to the wrist-plate as the case may be necessary to do.

If the engine has the half-moon, crab claw, or other gear which opens the valves toward the center of the cylinder, the manner of the adjustment will be the same, except that the "lap" on the steam valves will be shown when the line on the steam valve is nearer the end of the cylinder, and the "lead" when this line is nearer the center of the cylinder than the line on the valve chamber. The adjustment of the exhaust valves and the amount of "lap" and "lead" will be the same in either case.

To adjust the rods connecting the cut-off or tripping cams with the governor, have the governor at rest and the wrist-plate at one extreme of its travel. Then adjust the rod connecting with the cut-off cam on opposite steam valve so that the cam will clear the steel on the tail of the hook about $\frac{1}{32}$ inch. Turn the wrist-plate to the opposite extreme of travel and adjust the cam for the other valve in the same manner. To equalize the cut-off and test its correctness, hook the engine in and block the governor up about $1\frac{1}{4}$ inch, which will bring it to its average position when running. Then turn the engine slowly, in the direction in which it is to run, and note the distance the cross-head has traveled from its extreme position at dead center when the cut-off cam trips or detaches the steam valve. Continue to turn the engine beyond the other dead center and note the distance of cross-head from its extreme of travel when the valve drops. If the distance is the same as when the other valve dropped the cut-off is equal. If not, adjust either one or the other of the rods until the distances are the same.

By following these directions, the engine will do good work, but to know just what it is doing the engineer should use the indicator often. No engine room is complete without a good indicator, and no engineer can be well posted as to what his engine is doing and keep it in its best possible condition for good work without having an indicator and using it often. (See p. 68.)

THE DYNAMO.

Q. What is a Dynamo?

A. A Dynamo is a machine in which Electricity is gathered and forced out through wires for lighting, Electro-plating, etc.

Q. What does a Dynamo consist of?

A. A Dynamo consists of a field, frame, armature, commutator, brushes, brush holders, pins for the brush holders and a quadrant.

Q. What is meant by a field?

A. It means the magnets connected to the frame with bolts.

Q. What are magnets?

A. Magnets are iron cores, wound with insulated wire. These magnets are called electro-magnets because they become magnetic only when a current passes through the wire.

Q. How is the current generated?

A. By the rotary motion of the armature between the poles of the magnet.

Q. What does an armature consist of?

A. It consists of either a steel or iron shaft, around which insulated wire is wound, the shaft having a 6 or 8 inch bearing at each end.

Q. How is the current conducted to the lamps?

A. By means of brushes made out of copper strips or wires about 6 or 8 inches long, soldered together at one end and held on the commutator by means of brush holders made out of brass. These holders are on long pins, the pins are nutted to a quadrant and the quadrant is fastened to the frame.

Q. How many brushes are there generally, and where are they?

A. There are 2 and 4 brushes, two on one side of the commutator and two directly opposite, according to size of machine.

Q. What is a commutator?

A. A commutator is made out of segments of copper and segments of insulation.

Q. Can a commutator be taken off when worn out? A. Yes.

Q. How is it generally done?

A. By taking out the brushes, brush holders, the pins and the armature from the dynamo, then place the two ends of the shaft on wooden horses, mark the wires connecting the armature and commutator by attaching numbered tags (so as to place them, when the new commutator is put on) then disconnect the wires between the commutator and armature and take off the commutator from the shaft.

Q. How should a dynamo be looked after and run?

A. See that the machine is clean, journals cool, and that the proper speed is kept up; see that the brushes are directly opposite each other and that the quadrant and brushes are moved around on the commutatar according to the number of lights in use.

Q. How would you know when to move the quadrant?

A. By the sparking of the brushes on the commutator.

Q. What mainly causes the dynamo to flash or spark?

A. The brushes not being directly opposite through the diameter of the commutator, sometimes not enough pressure on the commutator, sometimes the brushes not far enough around on the commutator, also too much brush surface.

Sparkling at the brushes. Some styles of dynamos will spark at the brushes in spite of anything the attendant can do to prevent it, but many other styles of dynamos can be run with absolutely no sparks on the commutator. The first point to be attended to is to get your commutator perfectly smooth, or as near it as possible, with the means at your command, for if the commutator is not true you can not prevent it from sparking.

If you have a slide-rest, use it, and get your commutator round and true from end to end. If you have no slide-rest, a 16 in. bastard file will do nearly as well. Take the brushes and brush holders off, so that you may have plenty of room to work. Start the dynamo to turning very slowly. Hold a piece of chalk so near the commutator that it will mark all of the high spots. Move the chalk slowly from end to end of the commutator, so that all high places on the full length will be chalked. Stop the dynamo and amuse yourself filing off those parts that have been marked by the chalk. If you have noticed while the dynamo was turning about how much the commutator was "out," you can easily tell about how much you will have to file away to bring it true. File off all the places that have been marked, and then start up again slowly, and chalk it again. Repeat the chalking and filing until the commutator is round, and of the same size from end to end.

Next get a piece of shingle, thin board, or a piece of lathe even will do, and wrap a sheet of No. oo. *sand-paper* around it—never use emery paper or cloth—start the dynamo at a pretty lively speed, and smooth the commutator down with the sand-paper, holding the flat side against the work. It is not necessary to work it down to a *polished* surface, although it would be well if it were polished. Now that you have your commutator round and smooth— and it must be so smooth that there are none of the marks left on the commutator, for it was trouble that caused them, and if any be left they will certainly cause more trouble.

Now, that you know your commutator is in good shape, proceed to set your brushes, being certain that the points of opposite brushes are directly opposite through the diameter. The pressure put on the brushes need only be just sufficient to make *good contact*. It is not necessary to have much pressure to preserve good contact. Should the contact be too slight it will make itself known by a peculiar noise that is indescribable, being neither a snap, crack or pop, and yet might be called by either of these names. You may be sure that the noise will call your attention if you are anywhere near, and after you have once noticed it you will easily recognize it the next time. This noise and considerable sparking will always be present when the brushes do not press heavily enough upon the commutator.

If the brushes are not set with the points directly opposite, sparking will result.

If the brushes are set ahead of the neutral line or back of it they will spark.

When setting four brushes on a commutator that requires two brushes side by side, it is

sometimes difficult to get all four of them of
an equal length, or evenly divided on the commutator, one or more of them will spark more
or less. After rocking the brushes back and
forth a trifle to find the point of least sparking,
you can then tell by the color of the spark
whether the brush should be lengthened or
shortened. When the spark is of a decidedly
greenish color the brush is too short, but if the
spark appears to spatter and shows a reddish
hue, then you will find that the brush is too
long, or it is so worn that there is too much of
it in contact. By the way, you will find fully
as much, if not more, trouble arising from
having too much of the brush in contact, than
from having too little.

Cutting of Commutator, scratching and eat-
ing away of the segments, is mostly due to the
brushes having too much surface in contact, and
increase of pressure will wear away the commutator, and having too much of the face of
the brush in contact will cause an edge of the
segments to become eaten away, and if not attended to, they will, in a very short time, be-

come as rough and uneven as a corduroy road.

With the thicker style of brushes we have never found it necessary, even when running at full load, to have more than one-third of the full end surface of the brush in contact with the commutator, and further, we have found that if we allowed the brush to become so worn that even one-half of the end surface bore on the segments it would cause sparking.

To prevent filing the brushes every day (which would be wasteful), to keep them in the best of order, we found that they could, with great atvantage, be turned the other side up and allowed to wear in that way until the surface became to great. This resulted in getting more than twice the amount of work out of a brush than was possible by filing always from one side, or trimming the ends square as often as they became badly worn. If the commutator becomes very hot you will be quite sure to find that your brushes are badly worn.

Flat spots on the commutator, frequently explained by laying it to soft spots in the copper, we have always found to result from an entirely different cause. When the marks have the appearance of a blow from the pene of a hammer, it will generally be found to be caused by a loosely connected or badly soldered armature wire connection. A spot of this kind continues to grow larger until the cause of it is removed and the commutator dressed down smooth.

At the end of the segments a spark or stream of fire encircling the whole commutator will sometimes be noticed.

This *may* be caused by an accumulation of oil or copper-dust or dirt, that causes a short circuit, but it will generally be found that the insulation is charred or burned through at some place near where the spark is noticed, and if a careful examination of the armature wires are made you will find that a connection is loose or has very poor conductivity. Allowing the commutator to run hot will increase difficulties of this kind.

The Principle of the Dynamo compared with the Steam Pump.

We are often asked how can a dynamo be easily understood; the questions coming from engineers who have charge of electric lighting plants.

The whole thing may be compared, in its principles, to the working of a steam pump forcing water through a line of pipe of the same extent as the line wires. The dynamo (or pump) forces electricity instead of water. So long as the dynamo or pump works continuously the pipes or wires are filled with a current of water, or electricity, flowing in one direction; in other words; a continuous current. Thus we may say: that a certain number of pounds steam pressure is required to overcome the friction of the water in the pipes, so that so many cubic feet or gallons of water shall be delivered per minute, equally true we can say, so many volts are required to overcome the resistance of the wire, so that the current shall be delivered in so many amperes per minute. Hence, to simplify, we may say pounds of steam pressure = volts; the friction=resistance; the pipe=the wire; current =volume of water in motion, and amperes of electricity=gallons of water delivered at the end per minute. Every engineer knows that the larger the pipe the more gallons water per minute, and the less relative friction, so the larger the wire the more current can be carried and the less the resistance, relative to the number amperes delivered. The same analogy holds good in the opposite, for the smaller the pipe or wire, the greater the friction, or resistance. Every engineer who uses a steam pump or an injector, knows that there is some point to which, if his pipes were reduced in size, nearly or quite all his power (steam pressure) would be absorbed in friction. So electrically, our voltage may be largely consumed or absorbed by too small a wire; in either case—either the water or the electricity—the result of the work done is in both cases uniform and identical, viz.: A continuous current, and is the current that has been generally used for the production of light and power. The other current, which is largely employed in the generation of electrical power, viz.: the alternating current, differs essentially from that which we have described above, and infact our analogy to the working of a pump comes to an end. The current from an alternating dynamo, instead of flowing continuonsly and directly, is simply a vibratory movement, or a "back and forth flow." Here the supremacy of electricity as a power, or rather as a transmitter of power, comes in, for, returning to oue pump, should we at each alternate stroke of the pump reverse the direction of flow of the water, the entire power, or nearly all of it, would be absorbed by its weight, and tbe friction in the pipes. But electricity boing without weight, there is of course no loss by reversing its flow; indeed, the possibilities of application to useful service, dependent on the reversals, are of the greatest value. To clearly explain the action of the alternating system, we have to consider the requirements under which electricity does its most acceptable work.

Every engineer who is making electric lights knows that the most satisfactory results, *i. e.*, the best light, is obtained by using a dynamo and distributing system of as high voltage as possible, in conjunction with a lamp of low voltage. Here, then, we have two actually opposite conditions, which must be harmonized to produce a perfect result in their action, and which are plainly impossible in the continuous current system, which we have explained by the comparison to our pump; because it is evident, to renew the comparison; that if we are carrying a pressure (steam), and our line of pipes is calculated to deliver a certain amount of water per minute, if we throttle down at the delivery end, so as to deliver only $\frac{1}{10}$ or $\frac{1}{20}$ of the amount, we shall only be able to do so by reducing our pressure relatively, involving a great loss of efficiency, or incur the risk of destruction to the plant at some point.

Hence we are obliged to provide some appliance which shall intervene to convert the high voltage of the dynamo and circuit to the low voltage of the lamps. When such an appliance is used it is known as a converter system, and the use of an alternating current and converter system are mutually dependent on and necessary to each other.

This system can be compared to the engineer's system of steam heating in his building thus: Suppose he carries 75 lbs. boiler pressure, and the steam is carried into the building in one main pipe, and from that is distributed by risers, etc., to the different radiators in the building. It is evident that he has no use for full boiler pressure on the risers and radiators, as, even if they would stand it for a time, it would be no more effective for heating than a reduced pressure; hence he puts in a reducing valve in the steam main, between the boilers and risers.

So, then, the converter used in connection with an alternating current is exactly an electrical reducing valve, with a high pressure (voltage) on one side, and a low working pressure (voltage) on the other. Thus, by using this converter he may carry any voltage at the dynamo and primary circuit, reducing into the secondary, to conform to the amount of current required. Each current continuous or alternate, have especial fields to which they are adapted, and while both are extensively in use each has its peculiar adaptation.

Q. How do you understand the term "volt?"

A. The "volt" is a measure of electro—motive force, or original energy. Corresponding to the dynamic term "pressure," but not of "power." It is based on the product of one Daniell cell of a battery.

Q. How do you understand the term "ohm?"

A. The "ohm" is the measure of resistance, and compares to the dynamic term of "loss by transmission." It is based on the resistence offered by a copper wire .05 in. diameter, 250 ft. long; or a copper wire, 32 gauge, 10 ft. long.

Q. How do you understand the term "ampere?"

A. The "ampere," is the measure for current or what passes; the intensity, it may be called, and is comparable to the dynamic term of "power transmitted," or "effect." It is the residual force of one "volt" after passing through one "ohm" of resistance.

Q. How do you understand the term "coulomb?"

A. The "coulomb" is a measure of current, qualified by time; one ampere acting for one second of time, comparing in nature with the dynamic "foot-pound."

Q. How do you understand the term "watt?"

A. The "watt" is the unit for dynamic effect produced by electro-motive force or current. It equals 44.22 foot-pounds, or one 746 h. p.

Q How many "coulombs" in a "watt?"

A. There are 44.22 "coulombs."

Q. How many "watts" in an electrical h. p.?

A. There are 746 "watts" in a h. p.

Q. How many horse power will it take to run a 50 arc light dynamo. Each arc light equaling 45 "volts" and 8 "amperes" giving 1600 candle power to each light?

A. Multiply the "voltage" by the "amperes" then the number of lights lit, and divide by electrical h. p. which is 746 "watts." The answer will be the h. p. of engine required.

Receipts for Gold and Silver Plating.

Take a tablespoonful Cyanide of Gold and put it in a glass of water, to do gold plating.

———

All articles to be plated should be dipped in strong lye or diluted nitric acid, and rinsed off with soft water; then place the article to be plated in the glass that has the solution of either gold or silver, and take a couple of pieces of zinc 1 inch wide, and double to ½ in. wide, by 10 long, let it touch the article to be plated, and you will be surprised at the result. This answers for both.

To make solution of silver for plating:—Take silver and dissolve in a glass with little nitric acid, when the silver is dissolved then drop hydroloric acid in until the white precipitates, (silver chloride) ceases to fall, pour off the colored water after it has settled, and add soft water to it, then it is ready for use.

POINTS FOR ENGINEERS.

STEAM-PIPES, whether for power or for heating, should always pitch downward from the boiler, that the condensed water, etc., may have the same direction as the steam, or otherwise there will be trouble, unless the pipes are either very short or very large.

GLOBE valves should always be so placed in steam-pipes that their stems are very nearly horizontal, in order to prevent a heavy accumulation of condensed water in the pipes. Wherever a horizontal steam-pipe is reduced in size there should be a drip to avoid filling the larger pipe partially with condensed water.

———

IN order to make a rust joint that will stand heat and cold as well as rough usage, mix ten (10) parts of iron filings and three (3) parts of chloride of lime with enough water to make a paste. Put the mixture on the joint and bolt firmly; in twelve hours it will be set so that the iron will break sooner than the cement.

Rules for Calculating Speed and Sizes of Pulleys.

To find the size of driving pulley.

Multiply the diameter of the driven by the number of revolutions it shall make, and divide the answer by the revolutions of the driver per minute. The answer will be the diameter of the driver.

To find the diameter of the driven that shall make a given number of revolutions:

Multiply the diameter of the driver by its number of revolutions, and divide the answer by the number of revolutions of the driven. The answer will be the diameter of the driven.

To find the number of revolutions of the driven pulley:

Multiply the diameter of the driver by its number of revolutions, and divide by the diameter of the driven. The answer will be the number of revolutions or the driven.

Dia.	Cir.	Diam.	Cir.	Size	Area.	Size.	Area.
⅛	.3927	38	119.4	⅛	0.0123	38	1134.1
¼	.7854	39	122.5	¼	0.0491	39	1194.6
½	1.578	40	125.6	½	0.1963	40	1256.6
⅝	1.963	41	128.8	⅝	0.3068	41	1320.2
⅞	2.741	42	131.9	⅞	0.6013	42	1385.4
1	3.142	43	135.1	1	0.7854	43	1452.2
½	4.712	44	138.2	½	1.762	44	1520.5
2	6.283	45	141.4	2	3.142	45	1590.4
½	7.854	46	144.5	½	4.909	46	1661.9
3	9.425	47	147.6	3	7.068	47	1734.9
4	12.56	48	150.8	4	12.566	48	1809.6
5	15.71	49	153.9	5	19.635	49	1885.7
6	18.85	50	157.1	6	28.274	50	1963.5
7	21.99	51	160.2	7	38.484	51	2042.8
8	25.13	52	163.3	8	50.265	52	2123.7
9	28.27	53	166.5	9	63.617	53	2206.2
10	31.41	54	169.6	10	78.54	54	2290.2
11	34.55	55	172.8	11	95.03	55	2375.8
12	37.70	56	175.9	12	113.10	56	2463.0
13	40.84	57	179.1	13	132.73	57	2551.8
14	43.98	58	182.2	14	153.94	58	2642.1
15	47.12	59	185.3	15	176.71	59	2734.0
16	50.26	60	188.5	16	201.06	60	2827.4
17	53.40	61	191.6	17	226.98	61	2922.5
18	56.55	62	194.8	18	254.47	62	3019.1
19	59.69	63	197.9	19	283.53	63	3117.2
20	62.83	64	201.0	20	314.16	64	3217.0
21	65.97	65	204.2	21	346.36	65	3318.3
22	69.11	66	207.3	22	380.13	66	3421.2
23	72.25	67	210.5	23	415.48	67	3525.7
24	75.40	68	213.6	24	452.39	68	3631.7
25	78.54	69	216.7	25	490.87	69	3739.3
26	81.68	70	219.9	26	530.93	70	3848.5
27	84.82	71	223.0	27	572.56	71	3959.2
28	87.96	72	226.2	28	615.75	72	4071.5
29	91.10	73	229.3	29	660.52	73	4185.4
30	94.25	74	232.5	30	706.86	74	4300.8
31	97.39	75	235.6	31	754.77	75	4417.9
32	100.5	76	238.7	32	804.25	76	4536.5
33	103.6	77	241.9	33	855.30	77	4656.7
34	106.8	78	245.0	34	907.92	78	4778.4
35	109.9	79	248.2	35	962.11	79	4901.7
36	113.1	80	251.3	36	1017.9	80	5026.6
37	116.2	81	254.5	37	1075.2	81	5153.0

To find the Circum. of any Cir. see page 74.

To find the Area of any Circle or diameter see page 74.

INDEX.

RULES.

P.S.—None genuine without my address here. Books must be flexible, black, alligator, leatherette binding; red edge and round corners.

Send money in registered letter or postoffice order only. Price.....................

P. H. ZWICKER,

1607 Wash, Street, St, Louis. Mo.

AGENTS WANTED.